United States Nuclear Regulatory Commission

Protecting People and the Environment

NUREG/CR-7139
ANL-11/52

I0493908

Assessment of Current Test Methods for Post-LOCA Cladding Behavior

Office of Nuclear Regulatory Research

AVAILABILITY OF REFERENCE MATERIALS
IN NRC PUBLICATIONS

NRC Reference Material

As of November 1999, you may electronically access NUREG-series publications and other NRC records at NRC's Public Electronic Reading Room at http://www.nrc.gov/reading-rm.html. Publicly released records include, to name a few, NUREG-series publications; *Federal Register* notices; applicant, licensee, and vendor documents and correspondence; NRC correspondence and internal memoranda; bulletins and information notices; inspection and investigative reports; licensee event reports; and Commission papers and their attachments.

NRC publications in the NUREG series, NRC regulations, and Title 10, "Energy," in the *Code of Federal Regulations* may also be purchased from one of these two sources.
1. The Superintendent of Documents
   U.S. Government Printing Office Mail Stop SSOP
   Washington, DC 20402–0001
   Internet: bookstore.gpo.gov
   Telephone: 202-512-1800
   Fax: 202-512-2250
2. The National Technical Information Service
   Springfield, VA 22161–0002
   www.ntis.gov
   1–800–553–6847 or, locally, 703–605–6000

A single copy of each NRC draft report for comment is available free, to the extent of supply, upon written request as follows:
Address: U.S. Nuclear Regulatory Commission
          Office of Administration
          Publications Branch
          Washington, DC 20555-0001
E-mail: DISTRIBUTION.RESOURCE@NRC.GOV
Facsimile: 301–415–2289

Some publications in the NUREG series that are posted at NRC's Web site address http://www.nrc.gov/reading-rm/doc-collections/nuregs are updated periodically and may differ from the last printed version. Although references to material found on a Web site bear the date the material was accessed, the material available on the date cited may subsequently be removed from the site.

Non-NRC Reference Material

Documents available from public and special technical libraries include all open literature items, such as books, journal articles, transactions, *Federal Register* notices, Federal and State legislation, and congressional reports. Such documents as theses, dissertations, foreign reports and translations, and non-NRC conference proceedings may be purchased from their sponsoring organization.

Copies of industry codes and standards used in a substantive manner in the NRC regulatory process are maintained at—
    The NRC Technical Library
    Two White Flint North
    11545 Rockville Pike
    Rockville, MD 20852–2738

These standards are available in the library for reference use by the public. Codes and standards are usually copyrighted and may be purchased from the originating organization or, if they are American National Standards, from—
    American National Standards Institute
    11 West 42nd Street
    New York, NY 10036–8002
    www.ansi.org
    212–642–4900

Legally binding regulatory requirements are stated only in laws; NRC regulations; licenses, including technical specifications; or orders, not in NUREG-series publications. The views expressed in contractor-prepared publications in this series are not necessarily those of the NRC.

The NUREG series comprises (1) technical and administrative reports and books prepared by the staff (NUREG–XXXX) or agency contractors (NUREG/CR–XXXX), (2) proceedings of conferences (NUREG/CP–XXXX), (3) reports resulting from international agreements (NUREG/IA–XXXX), (4) brochures (NUREG/BR–XXXX), and (5) compilations of legal decisions and orders of the Commission and Atomic and Safety Licensing Boards and of Directors' decisions under Section 2.206 of NRC's regulations (NUREG–0750).

United States Nuclear Regulatory Commission

Protecting People and the Environment

NUREG/CR-7139
ANL-11/52

Assessment of Current Test Methods for Post-LOCA Cladding Behavior

Manuscript Completed: October 2011
Date Published: August 2012

Prepared by
M. C. Billone

Nuclear Engineering Division
Argonne National Laboratory
Argonne, IL 60439

H. H. Scott, NRC Project Manager

NRC Job Code V6199

Office of Nuclear Regulatory Research

ABSTRACT

Test methods to assess fuel-rod cladding behavior following a loss-of-coolant accident (LOCA) are compared and evaluated. For non-deformed cladding regions with uniform levels of hydrogen content and oxidation, the three-point bend test (3-PBT) is a very good test for ductility determination of as-fabricated and pre-hydrided cladding subjected to LOCA oxidation and quench. For irradiated cladding, the ring compression test (RCT) has clear advantages over the 3-PBT because the required sample length is only about one-tenth of that needed for the 3-PBT. Overall, the RCT is the best test method for generating ductility data for assessing the effects of irradiation and hydrogen pickup on embrittlement oxidation threshold. However, neither 3-PBTs nor RCTs are useful for evaluating the performance of ballooned and ruptured cladding with significant axial gradients in cladding geometry, oxidation level, and hydrogen content, as well as circumferential gradients in wall thickness and oxidation level within the rupture region. Partially restrained axial contraction tests are useful for determining the fracture/no-fracture boundary for ballooned, ruptured, oxidized, and quenched cladding as a function of hydrogen content and oxidation level. The four-point bend test (4-PBT) is best for determining three cladding performance metrics: maximum bending moment (measure of strength), failure energy (measure of toughness), and offset displacement (measure of plastic deformation).

FOREWORD

The purpose of this document is to support the technical basis for the choice of test methods to assess fuel-rod cladding behavior following a loss-of-coolant accident (LOCA). For non-deformed cladding regions with uniform levels of hydrogen content and oxidation, and for irradiated cladding, the ring compression test (RCT) has been selected. The four-point bend test (4-PBT) has been selected for evaluating the performance of ballooned and ruptured regions of a fuel rod in LOCA analysis. The technical basis is founded on the results of the NRC's LOCA research program, which was designed to measure the mechanical behavior of both non-deformed and ballooned and ruptured cladding following LOCA conditions. Integral LOCA tests were conducted at Argonne National Laboratory and Studsvik Laboratory in Sweden. The Japanese Atomic Energy Agency (JAEA) has also performed LOCA integral experiments. The results and observations of these experimental programs have been combined to develop considerations regarding the impact of oxidation and hydrogen content on the mechanical behavior of ballooned and ruptured cladding following LOCA conditions.

Brian W. Sheron, Director
Office of Nuclear Regulatory Research
U.S. Nuclear Regulatory Commission

CONTENTS

FIGURES

TABLES

EXECUTIVE SUMMARY

Current LOCA acceptance criteria in 10 CFR 50.46(b) limit the peak cladding temperature to 2200°F and the maximum oxidation level to 17% of the cladding wall thickness, assuming that all of the oxygen picked up by the cladding is in the form of ZrO_2. These criteria are intended to assure that cladding retains some ductility during and following quench. According to the 1973 Commission: "Our selection of the 2200°F limit results primarily from our belief that retention of ductility in the zircaloy is the best guarantee of its remaining intact during the hypothetical LOCA."

In the absence of a credible analysis of loads, cladding stresses, and cladding strains for a degraded LOCA core, there are no absolute metrics to determine how much ductility or strength would be needed to "guarantee" that fuel-rod cladding would maintain its geometry during and following LOCA quench. It is also not clear what impact severing of some fuel rods into two pieces would have on core coolability. Fragmentation of brittle fuel-rod cladding with very low fracture toughness would be more detrimental to core coolability than severing of rods into two pieces. However, it is well recognized that materials that can deform plastically prior to failure – and hence are classified as ductile – are more resistant to fragmentation than brittle materials with low fracture toughness. Materials that retain ductility can relax secondary thermal and differential-expansion stresses through plastic flow. Therefore, the intent to maintain ductility is beneficial even without adequate knowledge of LOCA loads.

Two test methods are currently being used as ductility screening tests for non-deformed cladding after exposure to LOCA oxidation and quench: ring compression tests (RCTs) and three-point bend tests (3-PBTs). Two other test methods are currently being used to assess the performance of ballooned and ruptured cladding following oxidation: post-quench four-point bend tests (4-PBTs) and partial-to-full axial restraint tests during quench. Other test methods have been employed to determine plastic stress-strain properties and ductility for cladding materials under conditions relevant to normal operation and reactivity-initiated accidents: axial tensile tests, hoop tensile tests, and tests with combined axial and hoop tensile stresses. These test methods are discussed and their relative advantages and disadvantages are summarized for determining cladding performance following LOCA oxidation.

Tests that require samples with end grips (e.g., axial tension) and/or machined gauge sections (e.g., axial and hoop tension) are not practical for LOCA-oxidized samples. Samples with high stiffness values relative to machine stiffness values are not desirable if one wants to compute failure energy, as well as ductility, from load-displacement curves. Tests that give load-displacement curves with a loading stiffness equal to the unloading stiffness are preferred if the simpler offset displacement method is used to determine ductility. Also, tests with smaller sample length are more practical for irradiated cladding because of limited availability and ease of defueling. Taking all these factors into account, the 3-PBT is the best choice for post-LOCA embrittlement determination of as-fabricated and pre-hydrided samples, which are not subjected to pre-oxidation deformation and are uniformly oxidized. Under the same conditions, the RCT is preferred for irradiated samples. However, it is desirable to have a single test method for as-fabricated, pre-hydrided, and irradiated cladding to eliminate systematic errors when comparing the behavior of these materials. For such comprehensive studies, the RCT is preferred.

Long pressurized cladding samples subjected to a LOCA transient can balloon and rupture, resulting in local areas with significant axial variation in diameter, wall thickness, hydrogen content, and oxidation level, as well as large circumferential variations in wall thickness and oxidation level within cross sections containing the rupture region. RCTs with samples sectioned from the balloon region are not appropriate because of the steep variation in outer diameter. Axial tension and bend tests are more appropriate for studying ballooned specimens.

To produce ballooned specimens for axial bend tests and axial-restraint tensile tests, Argonne (ANL) and the Japanese Atomic Energy Agency (JAEA) have performed LOCA integral experiments with lengths of as-fabricated and pre-hydrided cladding. In addition, JAEA has conducted experiments with defueled cladding sectioned from irradiated fuel rods.

Traditional axial tensile tests could be performed on post-LOCA integral samples, and pulling the sample to failure would give an accurate measure of failure load. However, if the load-displacement curve exhibited any offset displacement, it would not be clear if plastic displacement had occurred within the ballooned and ruptured region, just above and below this region where the temperature drops, or outside the middle region where the cladding is annealed, and lightly oxidized with negligible hydrogen pickup. The JAEA LOCA integral samples were not tested in the traditional way, but were fully or partially restrained from contracting during quench in the integral experiment. For samples that failed during quench, these tests generated data for axial failure load, failure location, and failure temperature. Although neither ductility nor failure energy could be determined, the JAEA test results do indicate that cladding can withstand high axial tensile loads without severing if the pre-oxidation hydrogen content and oxidation level are limited.

ANL has used four-point-bend tests (4-PBTs) to determine post-LOCA sample failure location, maximum bending moment (measure of strength), failure energy (measure of toughness), and offset displacement (measure of plastic deformation). The 4-PBT has a major advantage over the 3-PBT for ballooned and ruptured samples because 4-PBT loading does not bias the failure location. For 3-PBT loading, the bending moment varies from maximum at the load application point to zero at the two supports. For 4-PBT loading, the bending moment is constant between the load application points, which are spaced far enough apart to include the regions with non-uniform hydrogen content, oxidation level, and geometry. With analysis, the 4-PBT bending moment can be related to an equivalent axial tensile load.

In the ANL 4-PBTs, samples with rupture strains $\leq$32% failed at an axial location between the rupture edge and the hydrogen peak where the cladding was fully brittle. With a few exceptions, samples with $\geq$40% rupture strain failed in the rupture node location where the crack propagated from very brittle cladding (rupture tips) through brittle-ductile-transition material and through high-ductility cladding (back side of the balloon). The maximum bending moment and the failure energy were found to be strong functions of the oxidation that occurred after rupture of as-fabricated cladding. Equations A and B, respectively, show best-fit linear correlations for the maximum bending moment (M_{max}) and failure energy (E_{max}) as a function of calculated oxidation levels (CP-ECR, where CP refers to the use of the Cathcart-Pawel weight gain correlation) from 10% to 19%.

$$M_{max} = 13.92 - 1.073 \ (CP\text{-}ECR - 10\%), \ N{\bullet}m \tag{A}$$

$$E_{max} = 1.225 - 0.1236 \ (CP\text{-}ECR - 10\%), \ J \tag{B}$$

Between 19% and 23% (maximum tested) CP-ECR, both M_{max} and E_{max} are relatively small and constant. Equations A and B suggest that the oxidation level in the balloon region should be limited to enable the cladding to survive forces and moments during quench beyond the thermal stresses calculated for unrestrained cladding. At 17% CP-ECR, Eq. A predicts M_{max} = 6.4 N•m, which corresponds to an axial-restraint failure load of about 2600 N for samples with <31% rupture strain. This value is comparable to the maximum loads measured by JAEA (1200-2400 N) for ballooned, ruptured, oxidized and quenched samples that survived conditions of full restraint from contraction during quench from 700°C to ≈100°C.

Offset displacement can also be determined from 4-PBT load-displacement curves for ballooned, ruptured, oxidized, and quenched samples. However, for oxidation levels in the range of 10% to 18%, no offset strain was observed prior to the first significant crack severing >50% of the cross section containing the rupture flaw or all of the cross section within or outside the rupture region. These results apply to bend samples tested in the standard way with the brittle rupture tips subjected to axial tension and the ductile back region subjected to axial compression. Although it is clear that more than half the cladding cross section containing the rupture flaw had the capability of deforming plastically, most of the samples that failed in this region exhibited no offset displacement prior to cracking. Based on unflawed RCT samples, the ductile-to-brittle transition oxidation level is 19% for as-fabricated ZIRLO rings. Consistent with the RCT results, one 4-PBT sample at 18% CP-ECR subjected to "reverse" bending with the back region under axial tension exhibited a large offset displacement and did not even sever through 14-mm displacement. It was concluded that offset displacement is not a useful metric for assessing post-quench cladding performance of ballooned, ruptured and oxidized samples.

Based on the ANL results, the 4-PBT maximum bending moment and failure energy can be used to determine the resistance to fracturing and fragmentation of cladding subjected to ballooning, rupture, oxidation, and quench. The ANL data suggest that the current limits of 17% CP-ECR and 1204°C for fresh and very low-burnup fuel cladding are adequate for protecting the cladding during quench not only from fragmentation but also from severing into two pieces under a wide range of loading conditions. None of the samples tested to failure in the ANL program fragmented or failed in a "low-toughness" mode. Test samples severed at one or two locations. By contrast, glass and ceramic rods with much lower toughness than the post-LOCA samples sever into as many as 5 to 10 pieces when subjected to 4-PBT loading.

ACKNOWLEDGMENTS

The author would like to express his appreciation to the Office of Nuclear Regulatory Research (RES) for support and management of this program. As Project Manager, Harold Scott provided programmatic and technical guidance. John Voglewede and Michelle Flanagan provided technical guidance with regard to data needs relevant to licensing issues. John also performed a detailed technical review of this report.

The LOCA integral test data presented in this report were generated by ANL lead experimenters Yong Yan (now at Oak Ridge National Laboratory – ORNL) and Tatiana Burtseva. ANL technicians David McGann and Kevin Byrne are also acknowledged for their careful efforts in test setup, test conduct, and post-test measurements. Ralph Meyer provided valuable technical guidance for this test program and a detailed technical review of this report. He is also acknowledged for his in-depth assessment of ANL data and the comparison of these data to the international dataset for cladding behavior during and following a hypothetical loss-of-coolant accident.

The author would like to express his appreciation to international colleagues Jean-Christophe Brachet and Valerie Vandenberghe of the Commissariat a l'Energie Atomique (CEA) at Saclay and Fumihisa Nagase of the Japanese Atomic Energy Agency (JAEA) for sharing details and insights regarding test methodologies, data, and embrittlement mechanisms. Members of the international ad hoc Mechanical Properties Experts Group are also acknowledged for sharing test methodologies and data for ring tension tests. The group was formed in October 2000 and has included members from Anatech, CEA, Electricitie de France (EdF), Electric Power Research Institute (EPRI), Institute for Radiological Protection and Nuclear Safety (IRSN), JAEA, ORNL, Pennsylvania State University (PSU), Russian Research Center's Kurchatov Institute (RRC-KI), and Studsvik Nuclear AB.

ACRONYMS AND ABBREVIATIONS

3-PBT	3-point bend test
4-PBT	4-point bend test
AF	As-fabricated
ANL	Argonne National Laboratory
BJ	Baker-Just weight gain correlation
CEA	Commissariat a l'Energie Atomique
CFR	Code of Federal Regulations
cm	Centimeters
CP	Cathcart-Pawel weight gain correlation
°C	Degrees Celsius
ECR	Equivalent cladding reacted
EDC	Expansion due to compression
EdF	Electricitie de France
EPRI	Electric Power Research Institute
FEA	Finite Element Analysis
GWd/MTU	Gigawatt-days per metric tonne of uranium
ID	Inner diameter
IRSN	Institute for Radiological Protection and Nuclear Safety
J	Joule, measure of energy
JAEA	Japanese Atomic Energy Agency
kN	Kilonewton
LOCA	Loss-of-coolant accident
m	Meter, measure of length
mm	Millimeter
MPa	Megapascal, measure of pressure and stress
N	Newton, measure of force
NRC	Nuclear Regulatory Commission
OCZL	Out-of-cell oxidation test for ZIRLO
OD	Outer diameter
ORNL	Oak Ridge National Laboratory
PH	Pre-hydrided
ppm	(weight) parts per million
psig	Pounds per square inch gauge pressure
PSU	Pennsylvania State University
PWR	Pressurized water reactor
RCT	Ring compression test
RES	Office of Nuclear Regulatory Research
RRC-KI	Russian Research Center's Kurchatov Institute
RT	Room temperature
s	seconds
wppm	weight parts per million
Zry-2	Zircaloy-2
Zry-4	Zircaloy-4

SYMBOLS

a	Distance between applied load and support in 4-PBT
A	Cross-sectional area
A_{eq}	Equivalent cross-sectional area relative to the metal
A_g	Cross-sectional area for sample gauge section in axial tension test
A_m	Cross-sectional area of metal
A_{ox}	Cross-sectional area of oxide
$(A_{ox})_i$	Cross-sectional area of inner-surface oxide
$(A_{ox})_o$	Cross-sectional area of outer-surface oxide
d_{ap}	Permanent displacement at loading positions in 4-PBTs
D_i	Inner diameter
D_{mi}	Inner diameter of metal
D_{mo}	Outer diameter of metal
D_o	Outer diameter
d_p	Permanent displacement of RCT ring in loading direction
E	Young's modulus
E_g	Young's modulus for gauge material in axial tension test
E_m	Young's modulus for metal
E_{max}	Maximum applied energy in 4-PBT; equivalent to failure energy for failed samples
E_{ox}	Young's modulus for oxide
h	wall thickness
h_m	Wall thickness of metal
I	Area moment of inertia for bending
I_{eq}	Equivalent area moment of inertia relative to the metal
I_m	Area moment of inertia for metal
$(I_{ox})_i$	Area moment of inertia for inner-surface oxide
$(I_{ox})_o$	Area moment of inertia for outer-surface oxide
K	Stiffness; linearized loading slope of load-displacement curve
K_g	Stiffness of gauge section in response to axial load
K_m	Measured stiffness for RCTs and 4-PBTs; machine stiffness for axial tension tests
L	Length
L_g	Gauge length for axial tension test sample
L_n	Normalization length to convert bending displacement to axial strain
L_s	Distance between supports in 3-PBT and distance between applied loads in 4-PBT
M	Bending moment
M_{max}	Maximum bending moment
P	Load measured by load cell
P_{max}	Maximum load
R_{mo}	Outer radius of metal
T	Temperature
T_R	Rupture temperature
δ	Applied displacement
$\Delta C/C_m$	Circumferential strain at the mid-wall location
δ_e	Elastic displacement
δ_g	Elastic displacement of gauge section in axial tension test
δ_m	Elastic displacement of machine, including all components outside gauge length
$(\delta_{ox})_i$	Thickness of inner-surface oxide

1   BACKGROUND

Current LOCA acceptance criteria in 10 CFR 50.46(b) limit the peak cladding temperature to 2200°F and the maximum oxidation level to 17% of the cladding wall thickness, assuming that all of the oxygen picked up by the cladding is in the form of ZrO_2. These criteria are intended to assure that cladding retains some ductility during and following quench. According to the 1973 Commission: "Our selection of the 2200°F limit results primarily from our belief that retention of ductility in the zircaloy is the best guarantee of its remaining intact during the hypothetical LOCA."

In the absence of a credible analysis of loads, cladding stresses, and cladding strains for a degraded LOCA core, there are no absolute metrics to determine how much ductility or strength would be needed to "guarantee" that fuel-rod cladding would maintain its geometry during and following LOCA quench. It is also not clear what impact severing of some fuel rods into two pieces would have on core coolability. Fragmentation of very brittle fuel rod cladding would be more detrimental to core coolability than severing of rods into two pieces. It is well recognized that materials that can deform plastically – and hence are classified as ductile – are more resistant to fragmentation than brittle materials. In particular, materials that retain ductility can relax secondary thermal and differential-expansion stresses through plastic flow. Therefore, the intent to maintain ductility is beneficial even without adequate knowledge of LOCA loads.

The current LOCA acceptance criteria were based on ring-compression tests to assess oxidation and temperature limits at which embrittlement occurs. Ring-compression loading induces circumferential bending stresses in post-LOCA cladding samples. Under the load and above the support, the hoop stresses across the cladding wall vary from maximum tensile stress (inner surface) to maximum compressive (outer surface). At ±90° from the loading direction, the hoop stress distribution is reversed (i.e., compressive inner- and tensile outer-surface hoop stresses). However, ring-compression loading is not prototypical of other anticipated loads on the cladding during quench: axial stresses due to bending, axial stresses due to restricted thermal contraction of the cladding, and possible impact loading in the balloon region. As such, the ring-compression test should be viewed as a ductility screening test. Axial bending, axial tension, and hoop tension tests could also be used as ductility screening tests. These test methods are reviewed and assessed for post-LOCA mechanical tests of non-deformed and deformed-and-ruptured cladding samples following oxidation and quench.

2 TEST METHODS FOR NON-DEFORMED CLADDING

Several types of tests have been used as ductility screening tests to assess oxidation and temperature limits at which embrittlement occurs for cladding samples that are not deformed after exposure to LOCA conditions. Ring compression tests are described in Section 2.1. Three-point and four-point axial bend tests are discussed in Sections 2.2 and 2.3, respectively. Axial tensile and ring-stretch tests are discussed in Sections 2.4 and 2.5, respectively. The relative advantages and disadvantages of these test methods are summarized in Section 2.6.

2.1 Ring-compression tests (RCTs)

Argonne National Laboratory (ANL) has exposed as-fabricated, pre-hydrided, and high-burnup cladding samples to two-sided oxidation and quench. Short (about 8-mm-long) rings sectioned from these samples have been subjected to ring-compression tests to determine the ductile-to-brittle transition oxidation level as a function of hydrogen content. Results of such tests have been reported [1,2]. With the limited amount of high-burnup cladding available for such tests, there is a practical advantage in using ring compression testing in that a large number of tests can be performed with a relatively small inventory of cladding. Another practical advantage comes from lower gamma and beta-gamma worker dose accumulation per test for short samples.

Figure 1 shows a schematic of the ring cross section, loading, and displacement. The displacement (δ) is applied at a constant rate and the force (P) is measured by the load cell. All samples are compressed to failure, which may manifest itself as a single crack or multiple cracks through the sample wall and along the sample length. Ductility is determined from the permanent change (d_p) in ring diameter after unloading and/or by the offset displacement (δ_p) method. For two-sided-oxidized samples, cracking through the wall and along the length of the sample is equally probable at inner-cladding surface locations under the load (12 o'clock position) and above the support plate (6 o'clock position). The maximum bending moment and tensile hoop stress occur at these two locations. The bending moment and tensile hoop stress at positions $\pm90°$ (3 and 9 o'clock) from the loading axis are about 40% less within the elastic deformation regime. Also, the RCT stresses at $\pm90°$ from the loading axis vary from compressive at the inner cladding surface to tensile at the outer cladding surface. However, failure data are seldom generated at these locations because of the lower stress.

The ring in Fig. 1 acts like a spring in the elastic deformation regime with spring constant K such that P = K δ. For as-fabricated cladding alloys, the spring constant K can be expressed in terms of the Young's modulus E, length L, outer diameter D_o, and wall thickness h. For 8-mm-long, as-fabricated cladding rings tested at ANL, calculated stiffness values (K_c) ranged from 1440 N/mm (15×15 Zry-4 with D_o = 10.77 mm and h = 0.76 mm) at room temperature (RT) to 800 N/mm (17×17 Zry-4 with D_o = 9.50 mm and h = 0.57 mm) at 135°C. These values were calculated using the isotropic and recrystallized-annealed correlation for E given in MATPRO [3] for Zry-2 and Zry-4. Measured values (K_m) determined from load-displacement curves were generally within $\pm10\%$ of K_c for Zry-4 (see Fig. 2) and other cladding alloys (e.g., Zry-2, ZIRLO, and M5). The deviation was reduced to $\pm5\%$ when MATPRO anisotropic values for E_θ were used to determine K_c. MATPRO also has a correction for cold-work that would further reduce E (by about 5% for 25% CW). However, the as-fabricated cold-work levels for Zry-4 and ZIRLO alloys supplied to ANL by vendors were not provided with the material description. The observation that the measured loading slope agrees with the calculated loading slope is important as it indicates that the RCT load-displacement curve represents the response of the cladding ring independent of the compliance of the test machines used at ANL.

3

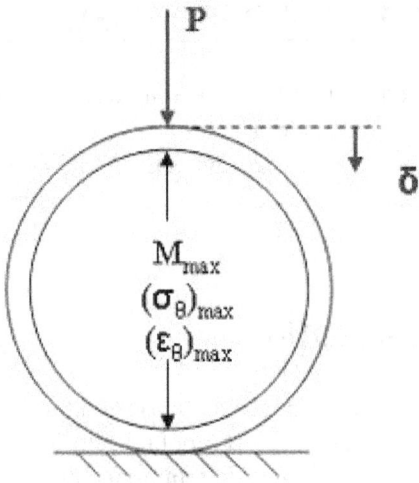

Figure 1. Schematic of ring-compression test sample and loading. The displacement (δ) rate is controlled and the response force (P) is measured. For two-sided-oxidized samples, failure occurs at one or both of the indicated locations. For one-sided-oxidized samples, failure may occur at locations ±90° from the loading direction.

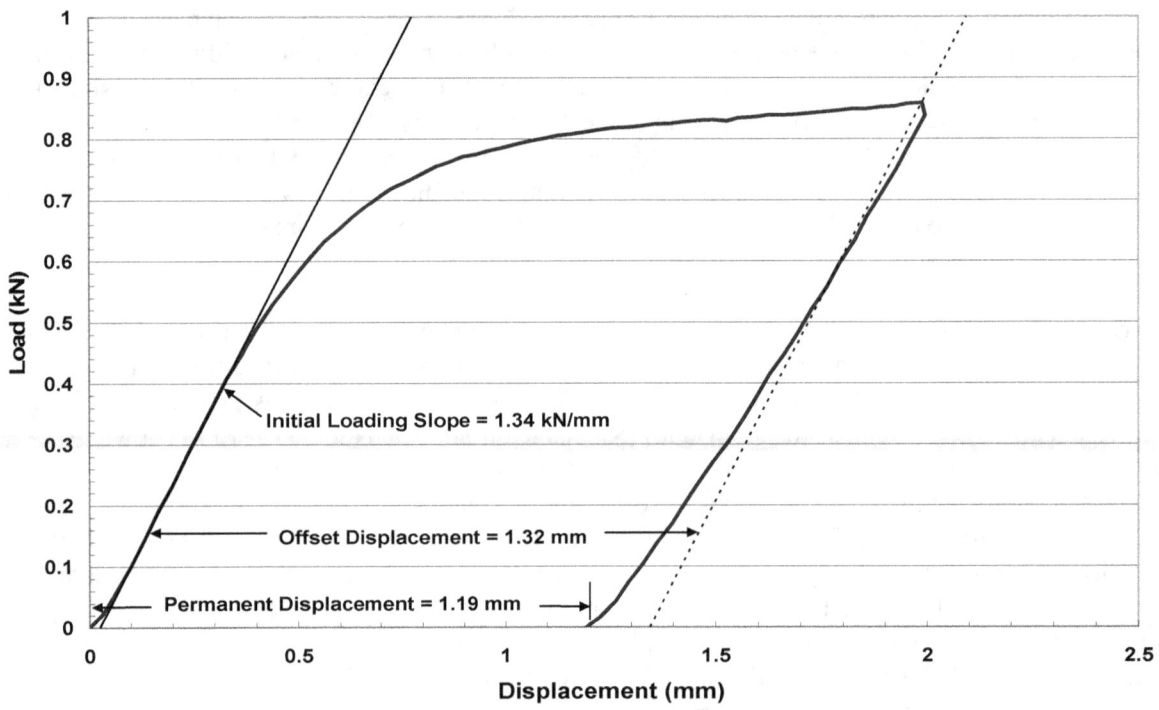

Figure 2. Load-displacement curve from a RT ring compression test with an 8-mm-long 15×15 Zry-4 cladding ring. The displacement rate was 2 mm/minute.

Rings with ductility will exhibit nonlinear load vs. displacement response beyond the elastic deformation regime. As shown in Fig. 2, the offset displacement (δ_p) was 1.32 mm while the permanent displacement (d_p) was 1.19 mm. The inherent error in δ_p comes from the assumption that the unloading slope (i.e., stiffness at the end of loading) is the same as the loading slope. The permanent displacement shown in Fig. 2 was in excellent agreement with the permanent change (pre-test minus post-test) in diameter measured directly from the sample. This subject is discussed in detail in Ref. 4. For post-LOCA rings tested to failure, the unloading slope of a ring just prior to failure cannot be determined. Thus, the loading slope must be used to determine δ_p. This approach is adequate to determine the ductile-to-brittle transition as long as the inherent error in δ_p is quantified.

It is common practice to normalize plastic displacement by the initial diameter (D_o) to determine relative plastic displacement. This is referred to as "strain" although it does not represent an average or maximum strain in the material. For example, the maximum elastic strain (ε_e) for an as-fabricated 17×17 Zry-4 ring is only about a third ($0.29\ \delta_e/D_o$) of the normalized elastic displacement. However, considering the complexity of oxidized and quenched rings, the determination of relative plastic displacement is accurate enough for a ductility screening test.

Because rings are compressed to failure in these tests, the measured permanent change in diameter also contains an error due to the unloading of a cracked ring. Based on experience and data trends, this error is <1% permanent strain. Therefore, rings exhibiting ≥1% permanent strain are classified as ductile, while rings that fail at <1% permanent strain are classified as brittle. For non-oxidized, as-fabricated cladding rings, the error between offset and permanent strain is negligible for small offset strain values (e.g., ≤3%). However, for oxidized-and-quenched cladding, the error increases with oxidation level. This deviation is shown in Fig. 3, which gives offset minus permanent strain (δ_p/D_o - d_p/D_o) as a function of oxidation level (in % CP-ECR). By setting the permanent strain (d_p/D_o) to the 1% ductility limit and adding it to the 1-σ upper bound in Fig. 3, the offset-strain ductility criterion can be expressed as:

$$\delta_p/D_o \geq 1.41\% + 0.1082\ \text{CP-ECR} \tag{1}$$

For new RCT data sets, the measured average offset strain (rounded to the nearest 0.1%) at a particular hydrogen content and oxidation level should be greater than or equal to the right-hand side of Eq. 1 (rounded to nearest 0.1%) for the material to be classified as ductile.

Ring compression tests have been conducted at many laboratories. However, ring support and loading surfaces (e.g., flat vs. curved), test conditions (e.g. test temperature and displacement rate), test procedures (e.g., stopping test after first significant load drop), and data interpretation (e.g., ductility based on total displacement vs. offset displacement vs. permanent displacement) vary considerably. Also, most laboratories did not search for the ductile-to-brittle-transition oxidation level as a function of hydrogen content. Thus, a one-to-one comparison could not be made between the dataset generated by ANL (see Fig. 4) and data generated at other laboratories.

Ductility and determination of ductile-to-brittle transition oxidation level have been the focus of the material presented on ring-compression testing. However, it has also been demonstrated that the load-displacement results are independent of machine compliance. Thus, one could determine failure energy (normalized to length) of the ring from the test results.

At their research center in Saclay, the French Atomic Energy Commission (CEA) performed ring-compression tests with as-fabricated and pre-hydrided Zry-4 and M5 cladding samples oxidized at 1200°C, as well as at lower and higher temperatures. RCT results for 1100°C and 1200°C oxidation

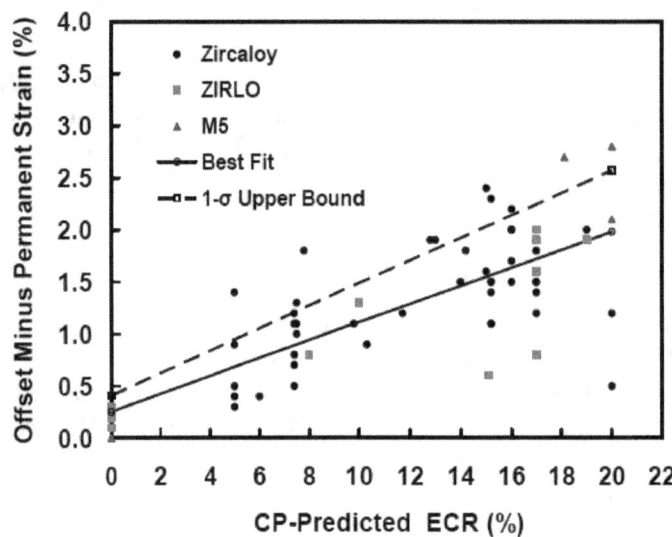

Figure 3. Difference between offset and permanent strains as a function of oxidation level for as-fabricated, pre-hydrided, and high-burnup cladding alloys oxidized at 1200°C and ring-compressed at 0.0333 mm/s. The dataset includes: 15×15 Zry-4; 17×17 Zry-4, M5 and ZIRLO; and 10×10 Zry-2 compressed at RT and 135°C to permanent strains of 1.0-2.3%.

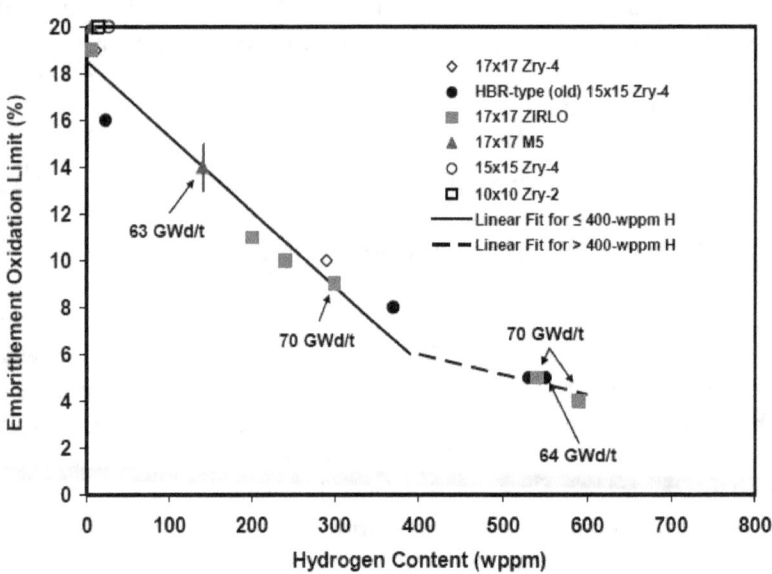

Figure 4. Embrittlement oxidation limit (CP-ECR in %) as a function of metal hydrogen content for as-fabricated, pre-hydrided, and high-burnup cladding alloys oxidized at a peak temperature of 1200°C, quenched at 800°C, and ring-compressed at 135°C. Oxidation values above data points and/or trend lines resulted in brittle behavior. For hydrogen contents in the range of 540-600 wppm, peak cladding temperatures (1130-1180°C) occurred during the heating ramp at <1200°C.

6

temperatures are presented in Refs. 5 and 6. Differences between CEA and ANL oxidation-quench tests include: heats (i.e., fabrication lots) of Zry-4 and M5 samples; 150-mm vs. 25-mm sample length; one-sided vs. two-sided oxidation, fast (20°C/s) vs. slow (2-3°C/s) temperature ramps within 100°C of oxidation hold temperature; and direct quench at oxidation temperature vs. about 13°C/s cooling rate from oxidation temperature to 800°C quench temperature. However, CEA also ran some tests with cooling rates of 0.4°C/s and 10°C/s from 1200°C to 800°C with quench temperatures of 800°C, 700°C, and 600°C [5,6]. ANL conducted a few tests with quench temperatures of 700°C and 600°C [1]. With regard to RCTs, differences between CEA and ANL include: 10-mm vs. 8-mm sample length; curved vs. straight support plate; 0.5 vs. 2 mm/minute displacement rate; and fixed displacement resulting in multiple cracks vs. stopping tests at the first significant load drop. The CEA approach yielded valuable data for offset strain, but data for permanent strain could not be determined for CEA test samples because of multiple cracks or wide single cracks due to continued displacement beyond the point at which the first through-wall crack developed.

For non-oxidized, as-fabricated 17×17 Zry-4, the RCT stiffness for CEA rings is calculated to be 1000 N/mm at 135°C. As CEA has not published load-displacement curves for such material, it cannot be determined if the slope of the load-displacement curves agrees with the calculated ring stiffness. Such agreement does not affect the determination of offset strain. However, if machine and fixture compliance is low enough, the measured loading slope could be lower than calculated for the ring, and the failure energy determined from the load-displacement curve could be higher than the failure energy for the ring. The stiffness of the CEA machine loading and support fixtures would have to be >10 kN/mm for the measured loading stiffness to be within 10% of the calculated ring loading stiffness.

In general, ANL and CEA RCT results were consistent in terms of ductile vs. brittle behavior for cases in which the pre-oxidation hydrogen content, oxidation level, cooling rate, and quench temperature were similar (see Fig. 7 in Ref. 6).

2.2 Three-point-bend tests (3-PBTs)

Figure 5 is a schematic of the three-point bend test (3-PBT). The applied load at the mid-span distance between supports results in a bending moment distribution that is maximum ($P \cdot L_s/4$) at the applied loading position ($L_s/2$) and decreases linearly to zero from the loading position to the two support positions. The top and bottom surfaces in Fig. 5 are subjected to axial compressive and tensile stresses, respectively. Thus, the maximum axial tensile stress occurs at the axial location at which the load is applied. With the load applied to the top surface, the maximum axial tensile stress occurs at the bottom surface.

CEA conducted both ring-compression tests (RCTs) and three-point-bend tests (3-PBTs) using as-fabricated and pre-hydrided cladding samples oxidized (outer surface only) at 1200°C and either quenched at 1200°C, slow cooled to lower temperatures and quenched, or cooled at 10°C/s to lower temperatures and quenched [5,6]. For the CEA bend samples, the span L_s between supports was 80 mm and the total sample length was 90 mm. The controlled displacement rate at the loading position was 0.5 mm/minute, the same as used in the CEA RCTs. As cladding inventory and dose rate are not issues for these materials, the use of long samples for 3-PBTs is not a disadvantage. Also, the 3-PBT has an advantage over the RCT inasmuch as loading and unloading stiffness values in bend tests should be essentially the same whereas they are different for RCT samples. Another advantage of the 3-PBT is that

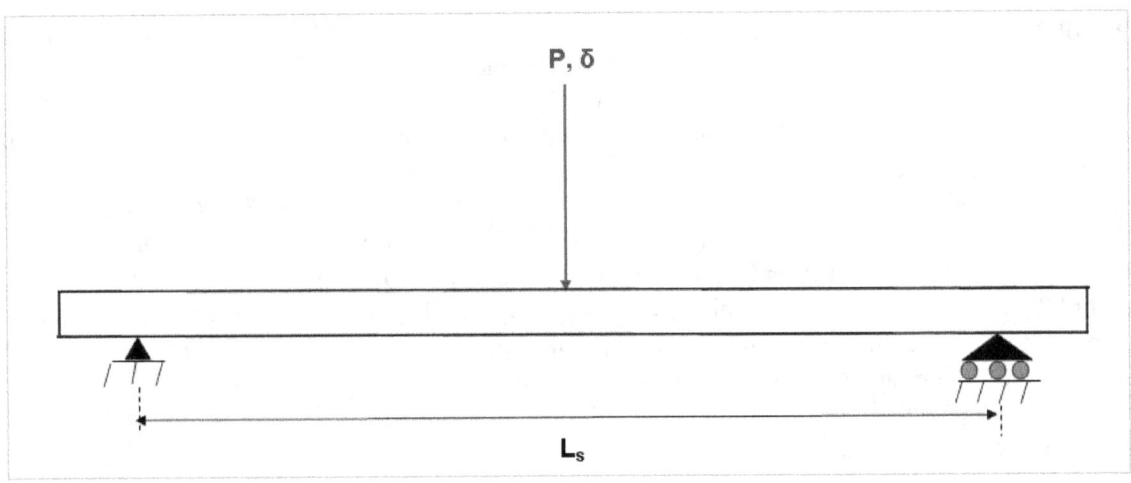

Figure 5. Schematic of the 3-PBT loading and support locations. The total length of the sample is L and the distance between supports is the span L$_s$.

failure is characterized by a single crack through the cross section at the loading position. In the absence of multiple cracks that form in RCT samples, it is straightforward to determine offset-displacement values from load-displacement curves. One disadvantage of the 3-PBT is that permanent displacement cannot be determined from pre- and post-test measurements for samples that sever into two pieces.

Because axial stresses are likely to occur during LOCA quench, the 3-PBT loading, which induces axial tensile stresses on the convex surface (bottom surface in Fig. 5) of the bent sample, is LOCA-relevant. However, CEA concluded that no significant difference exists between embrittlement oxidation levels determined from ring tests (hoop bending stress) and axial bend tests (axial bending stress). This implies that if the prior-beta layer is brittle in the hoop direction it will also be brittle in the axial direction [5,6]. However, it is difficult to compare displacement and strain data sets directly because the length-normalization factor to convert displacement to strain has not been determined for the 3-PBT sample. Also, unlike the RCT, the offset displacement and strain uncertainty levels have not been determined for the 3-PBT. The displacement uncertainty is estimated to be low but non-zero. The uncertainty in offset-displacement measurement and the length-normalization factor are estimated in this section for CEA 3-PBT results to allow a direct comparison of these results with CEA RCT results.

CEA RCT and 3-PBT results in Refs. 5 and 6 are for pre-hydrided ($\approx$600-wppm H) 17×17 Zry-4 oxidized to about 3% to 6% measured ECR and ring-compression tested at 135°C. The displacement rate for both test types was 0.5 mm/minute. There is no *a priori* reason why the same displacement rate would give the same material strain rate for these tests. However, for as-fabricated, non-oxidized cladding, the maximum elastic strain rate for the CEA 3-PBT sample is about 60% of the strain rate for the RCT sample at the 12 and 6 o'clock positions and about equal to the maximum RCT elastic strain rate at the 3 and 9 o'clock positions. Thus, CEA's use of the same displacement rate for both types of tests is justified.

The calculated stiffness for the 80-mm-span CEA bend sample loaded at mid-span is 1300 N/mm at 135°C for as-fabricated, non-oxidized 17×17 Zry-4 with 9.50-mm OD and 0.57-mm wall thickness. This stiffness is 30% higher than the stiffness for RCT samples sectioned from the same material. Stiffness values for oxidized samples may be higher due to the higher Young's modulus for the oxide as compared to the metal. On the other hand, the addition of 600-wppm of hydrogen would tend to lower the Young's modulus of the cladding metal.

The 3-PBT loading biases the failure location to coincide with the loading location, at which the bending moment, axial tensile stress, and axial tensile strain all reach their maximum value. The bending moment decreases linearly from the mid-span to the supports. For uniform cladding geometry, hydrogen content and oxidation level, failure should occur at the loading location.

An advantage of the 3-PBT is that the displacement is measured at the probable failure location. Thus, the load-displacement curve can be used to determine directly the offset displacement at the failure location. For the axial bend sample, the unloading slope is essentially equal to the loading slope. Thus, unlike the ring-compression results, there is no significant error in the determination of offset displacement using the loading slope to "unload" the sample prior to severing of the cross section.

Unlike the ring-compression test, there is no obvious choice of a dimension to normalize the 3-PBT offset displacement to calculate relative plastic displacement or strain. Intuitively, it should be the span length or some fraction of the span length. CEA could perform a finite-element analysis (FEA) to calculate maximum tensile plastic strain as a function of displacement for as-fabricated, non-oxidized cladding, as well as benchmark 3-PBTs with such cladding. In the absence of such an analysis, ANL has performed an elastic analysis to determine the appropriate length (L_n) to convert applied displacement to strain equivalent to the maximum axial tensile strain. This length was determined to be: $L_n = L_s^2/(12\,D_o)$, where L_s is the cladding sample length between supports and D_o is the cladding sample OD. For the CEA sample geometry, $L_n = 56$ mm, which is considerably larger than the 9.50 mm (sample OD) that CEA used to normalize 3-PBT offset displacements to determine offset strain (Fig. 5b in Ref. 6). Based on ANL FEA and benchmark data for the 4-PBT (see next section), the normalization length for offset displacement is less than the length for elastic displacement. As 56 mm is closer to the half span (40 mm) than the full span (80 mm) for the CEA tests, a preliminary recommendation is that $L_n = (L_s/2) = 40$ mm be used to convert CEA 3-PBT offset displacement to offset strain. However, ANL and CEA agree that there is no clean and simple way to compare RCT and 3-PBT results in terms of offset strain because the 3-PBT normalization length changes with displacement and oxidation level.

In terms of offset displacement uncertainty for bend tests, ANL has conducted 4-PBT loading-unloading-reloading experiments to determine loading and unloading slopes as a function of plastic displacement. Within the range of interest, there was no significant difference. Thus, for homogeneous material, one would expect no inherent error in the determination of offset displacement. However for non-homogeneous material, such as oxidized cladding, there may be an uncertainty of about 0.1 mm in the determination of offset displacement. In Ref. 6 (see Fig. 6), CEA has published 3-PBT load-displacement curves for 600-wppm-H Zry-4 oxidized to about 3% to 6% measured ECR, cooled at different rates, and quenched at different temperatures. The results range from near-zero (3% ECR oxidized and quenched at 1200°C) offset displacement to values as high as 2.7 mm (6% ECR slow cooled to 600°C and quenched). Based on fractography using scanning electron microscopy (SEM) images, it is clear that the sample with near-zero 3-PBT offset displacement is brittle. Yet the 3-PBT offset displacement was measured by ANL to be 0.06 mm based on the load displacement curve given in Fig. 6 of Ref. 6. Thus, it would be prudent to place an uncertainty limit of about 0.1 mm on 3-PBT offset displacement determination and assessment of plastic deformation. The corresponding RCT limit for samples oxidized to 3-6% ECR is 0.19 mm (see Fig. 3). These limits are shown in Fig. 6a for CEA RCT offset displacement vs. 3-PBT offset displacement (note: values were calculated from data plotted in Figs. 5b and 7b of Ref. 6). Eight of the data points plotted showed excellent agreement with respect to ductile vs. brittle behavior determined by both test methods: 5 ductile data points and 3 brittle data points. Two of the data points were brittle based on the RCT criterion (<0.19 mm) and ductile based on the 3-PBT criterion (≥0.1 mm). The agreement is quite good given the different stress states (hoop vs. axial) induced by the two test methods.

The offset displacement data shown in Fig. 6a were normalized by 9.50 mm for RCT and 40 mm for 3-PBT to generate the offset strain data shown in Fig. 6b. The conclusions are the same with regard to the very good agreement between the two test methods in distinguishing ductile vs. brittle behavior.

CEA also plotted "spent energy" or failure energy in Figs. 5d and 7d of Ref. 6. Depending on the effective stiffness of the CEA 3-PBT fixtures and tensile machine, these energy values may overestimate the energy stored in the samples prior to failure. With an ANL-calculated sample stiffness of 1300 N/mm, the machine stiffness would have to be >12.6 kN/mm in order for the energy determined from the linear part of the 3-PBT load-displacement curve to be within 10% of the elastic energy stored in the sample.

2.3 Four-point-bend Tests (4-PBTs)

The four-point-bend test (4-PBT, see Fig. 7) has advantages over the 3-PBT if there is axial variation in cladding geometry, hydrogen content, and/or oxidation level. A uniform bending moment is applied along the span L_s between the two loading points. The sample will fail at its weakest location or locations. If the sample is very uniform and brittle (e.g., glass or ceramic rod), then simultaneous failure at multiple locations is likely to occur.

For the ANL loading and support fixtures, L_s = 150 mm, a = 50 mm, and the distance between supports is 250 mm. The overhang regions beyond the supports are each 25 mm, but this length does not affect the loading. Thus, the uniform bending moment (in N•m) is given in terms of the measured load (in N) by M = (0.025 m) P. This bending moment results in maximum tensile stresses and strains at the bottom surface of the sample shown in Fig. 7, which would be convex during bending. The stresses and strains transition from maximum axial tension at the bottom surface to zero at the cross-section neutral axis to maximum axial compression at the top surface. The maximum stresses and strains are uniform within the span L_s for uniform geometry and material properties.

The 4-PBT sample is much more flexible than the RCT sample or the CEA 3-PBT sample. Benchmark tests have been conducted at RT and 135°C with 15×15 Zry-4 (D_o = 10.77 mm and h = 0.76 mm) and 17×17 M5 (D_o = 9.50 mm and h = 0.57 mm). Calculated stiffness values varied from 246 N/mm (15×15 Zry-4 at RT) to 121 N/mm (17×17 M5 at 135°C). Measured values (248±2 N/mm and 121 ±1 N/m, respectively) were in excellent agreement with calculated values. FEA calculations were also performed for 15×15 and 17×17 Zry-4 based on ANL-measured tensile properties for 15×15 Zry-4. The FEA-calculated load-displacement curves for the 15×15 Zry-4 benchmark tests at RT were in excellent agreement with the measured load-displacement curves.

Based on a comparison of the maximum elastic strain vs. displacement (δ) for the 4-PBT sample and the normalized RCT displacement (δ/D_o), the displacement rate for the 4-PBT was increased from 2 mm/minute to 2 mm/s. This was later reduced to 1 mm/s based on a direct comparison between calculated maximum elastic strain values for the RCT sample (ε_θ = 0.0305 mm^{-1} δ) and the 4-PBT sample (ε_z = 0.00104 mm^{-1} δ).

For one of the RT benchmark tests, 15×15 Zry-4 was subjected to a displacement of 10 mm at the loading points, which was large enough to induce plastic flow. The load-displacement curve for this test is shown in Fig. 8a. Although loads were not measured during unloading, the subsequent reloading stiffness was measured and found to be within 1% of the original loading stiffness. From this it was concluded that: the unloading stiffness was also the same and there would be no inherent error in determining 4-PBT offset displacement as there is for the RCT.

10

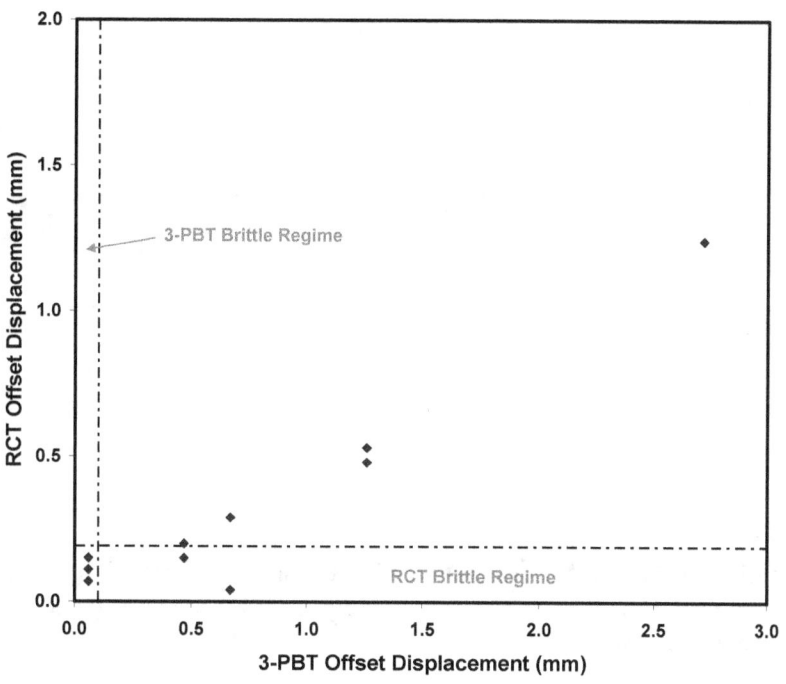

(a) Offset displacement

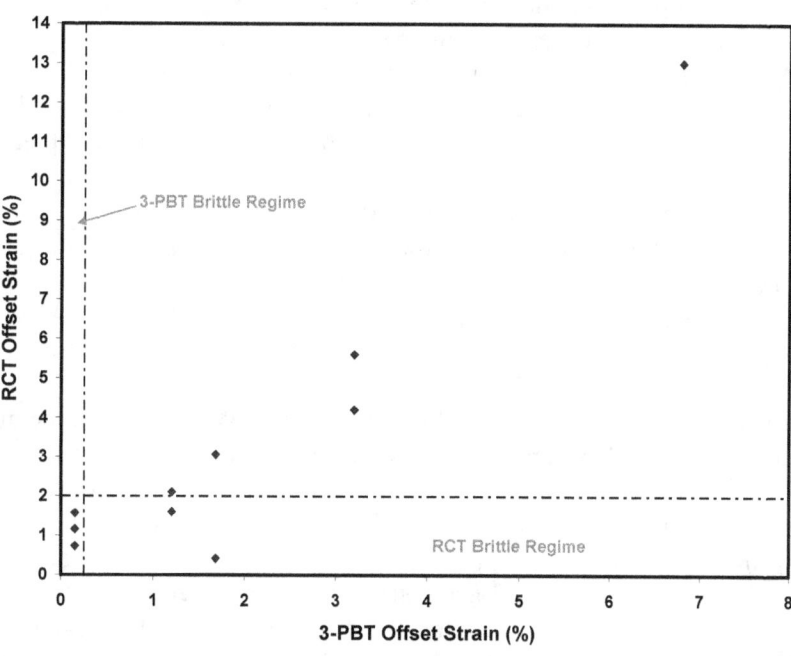

(b) Offset strain

Figure 6. CEA RCT and 3-PBT data generated at 135°C for pre-hydrided (≈600 wppm) 17×17 Zry-4 cladding samples oxidized at 1200°C to about 3 to 6% measured ECR and cooled at either ≤0.4°C/s or 10°C/s to quench temperatures of 800°C or 600°C or cooled slowly with no quench: (a) offset displacement and (b) offset strain.

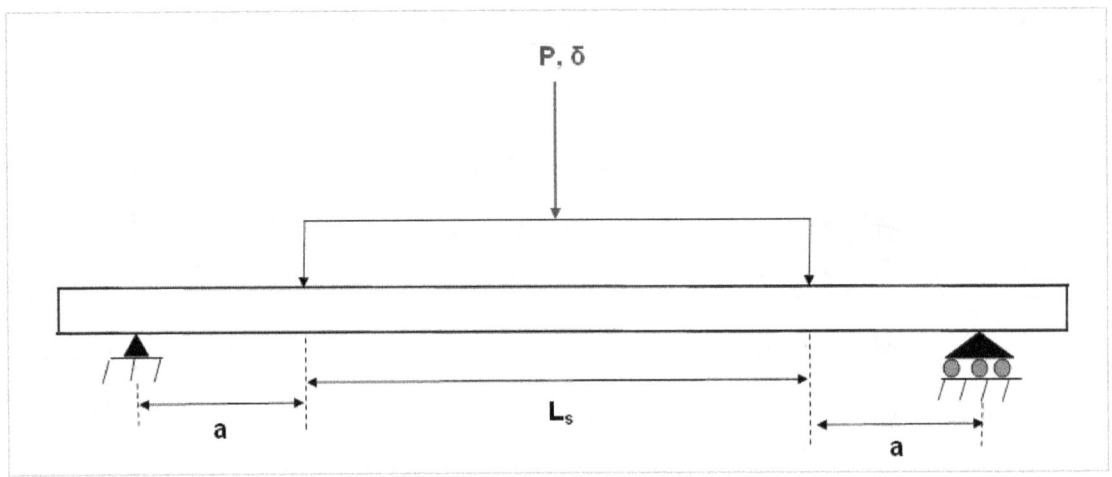

Figure 7. Schematic of four-point-bend test (4-PBT) loading. The bending moment [M = (P/2) a] is constant within the span L_s. The displacement (δ) rate is controlled and the force (P) is measured by the load cell.

From Fig. 8a, the offset displacement was determined to be δ_p = 1.9 mm at the sample loading points. The post-test deformed sample is shown in Fig. 8b. Accurate determination of the permanent displacement at the loading points requires enlargement of Fig. 8b. The approximate value of permanent displacement at the loading points is d_{ap} = 2.4 mm based on a reduced-size (0.828:1) photograph and d_{ap} = 2.1 mm based on an enlarged (1.40:1) photograph. The 2.1-mm permanent displacement is in good agreement with the 1.9-mm offset displacement. It should be noted that permanent displacement cannot be measured accurately for oxidized samples tested to failure (i.e., severed into 2 pieces).

The permanent plastic strain can, in principle, be determined from the radius of curvature of the deformed sample. From beam theory, the relationship between the maximum plastic strain (ε_p) and the radius of curvature (ρ_p) of the neutral axis is given by:

$$\varepsilon_p = (D_o/2)/\rho_p \qquad (2)$$

Based on an enlarged view of Fig. 8b, ρ_p = 1560 mm and ε_p = 0.35%. However, the measurement of radius of curvature is tedious and cannot be done accurately for an oxidized sample tested to failure.

A third approach to determining plastic strain is to compare the offset displacement at several displacements beyond the elastic regime to FEA-calculated maximum plastic strains. The calculation was performed for both 15×15 and 17×17 Zry-4 geometries. The experimental results in Fig. 8a were used to determine offset displacement vs. total displacement for 15×15 Zry-4. Analytical load-displacement results were used for 17×17 Zry-4. The results are given in Table 1 for offset displacement, calculated maximum plastic strain, and the normalization length (L_n) needed to convert offset displacement to plastic strain. For small plastic strains in the range of 0.16% to 0.30%, the approximate value for L_n is 250 mm, which is the distance between supports in Fig. 7.

From a practical perspective, it is sufficient to use offset displacement as the metric for plastic deformation for the ANL 4-PBTs. Mathematically, δ_p > 0 implies plastic flow and ductility. Taking into

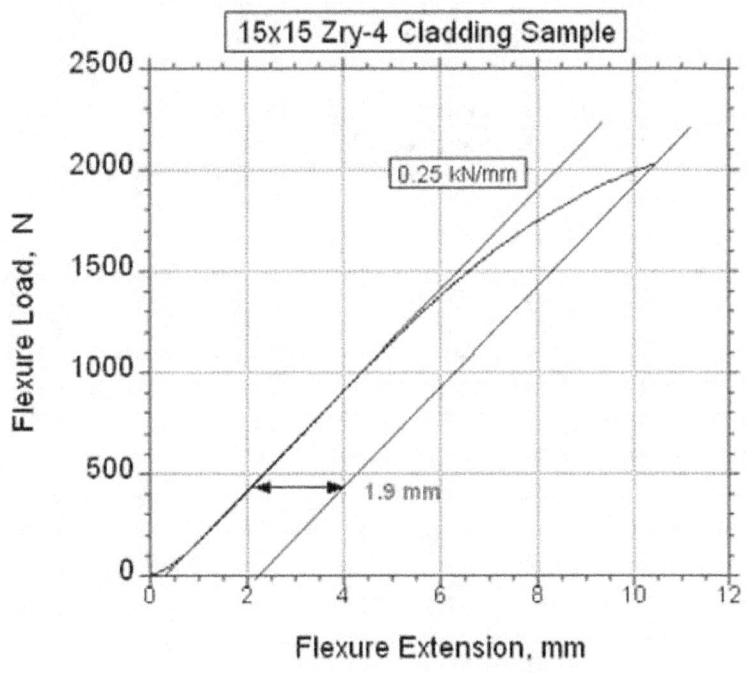

(a) Load-displacement curve

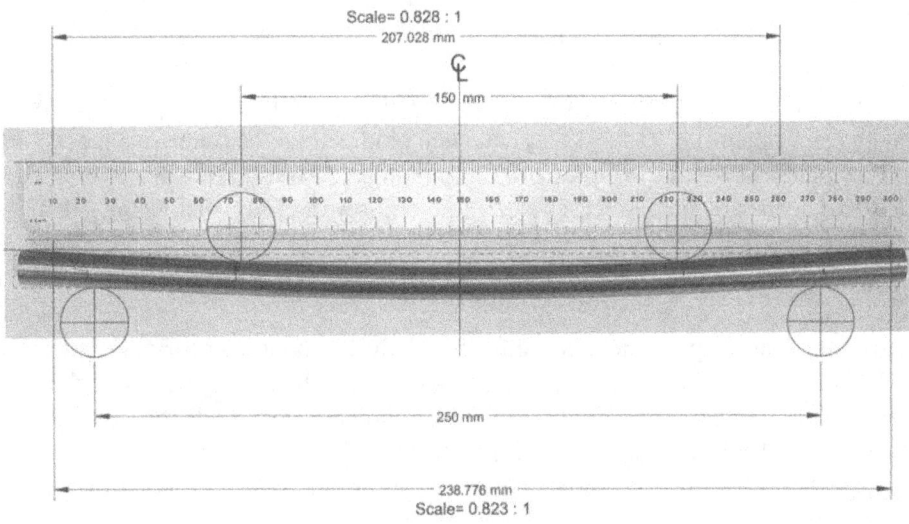

(b) Post-bend-test sample

Figure 8.   Results for 4-PBT conducted at RT with an as-fabricated 15×15 Zry-4 sample: (a) total load (P) vs. displacement (δ) at loading points and (b) deformed shape of post-bend sample.   Scale of 0.828:1 refers to original drawing size and is not relevant to Fig. 8b.

account measurement errors, it would be prudent to set a lower limit on offset displacement in the determination of ductile vs. brittle behavior: $\delta_p \geq 0.1$ mm implies ductile behavior. In addition to plastic deformation, material strength (as measured by the maximum bending moment) and failure energy (area under load-displacement curve) can be determined.

Table 1. **Determination of normalization length (L_n) to convert offset displacement to maximum plastic strain for ANL 4-PBT samples with uniform geometry and material properties along the axial direction. Results are for as-fabricated Zry-4 cladding samples subjected to 4-PBTs (see Fig. 7) at RT.**

Cladding Geometry	Offset Displacement, mm	Maximum Tensile Plastic Strain, %	Normalization Length, mm
17×17	0.31	0.16	194
15×15	0.50	0.20	250
15×15	0.90	0.30	300
17×17	1.56	0.38	410
15×15	1.90	0.50	380

Additional benchmark tests were conducted with 17×17 M5 cladding subjected to 4-PBTs at RT and 135°C. The displacement rate was 2 mm/s for these tests. Measured stiffness values were 131±1 N/mm at RT and 121±1 N/mm at 135°C. The measured values were in excellent agreement with the calculated values of 128 N/mm and 121 N/mm, respectively.

Four-point-bend tests were also performed with oxidized and quenched samples. One such test was performed with a specimen (OCZL#24) that had been oxidized on the outer surface at 1200°C to 17% CP-ECR and quenched at 800°C. Similar ZIRLO samples that were oxidized (two-sided) at 1200°C to 17% CP-ECR had been subjected to RCTs and exhibited low ductility. Thus, the 4-PBT should also demonstrate low ductility if the two test methods yield consistent results with regard to the ductile-to-brittle transition oxidation level.

Figure 9a shows the load-displacement curve for this 4-PBT conducted at 135°C and a displacement rate of 1 mm/s. The measured offset displacement was 0.4 mm. When compared with the 0.1 mm threshold offset displacement discussed above, this 4-PBT result indicates low ductility. The results also demonstrate that the 4-PBT is sensitive enough to distinguish ductile from brittle behavior even near the transition oxidation level. Figure 9b shows the post-test sample, which failed at two locations (±38 mm from axial mid-span location).

The middle region in Fig. 9b was sectioned into one sample (4-mm-long) for metallographic examination and three samples (each 8-mm-long) for ring compression testing. As the sample had already experienced some axial plastic flow, the expectation was that compressed rings would exhibit near brittle behavior. Load-displacement curves for these three rings are shown in Figs. 10-12. The results of the ring-compression tests are summarized in Table 2. Two of the post-bend rings did exhibit brittle behavior and on average the rings were brittle: 1.5±0.4% offset and 0.9±0.3% permanent strains.

It is worth noting that the average of the measured loading stiffness values for the compressed rings was 1000 N/mm. For non-oxidized, as-fabricated ZIRLO, compressed at 135°C, the measured stiffness values were 830 N/mm [4]. Thus, the presence of the outer-surface oxide layer increased the stiffness by about 20%. This is consistent with the uniform oxidation of the rings in this region of the test sample.

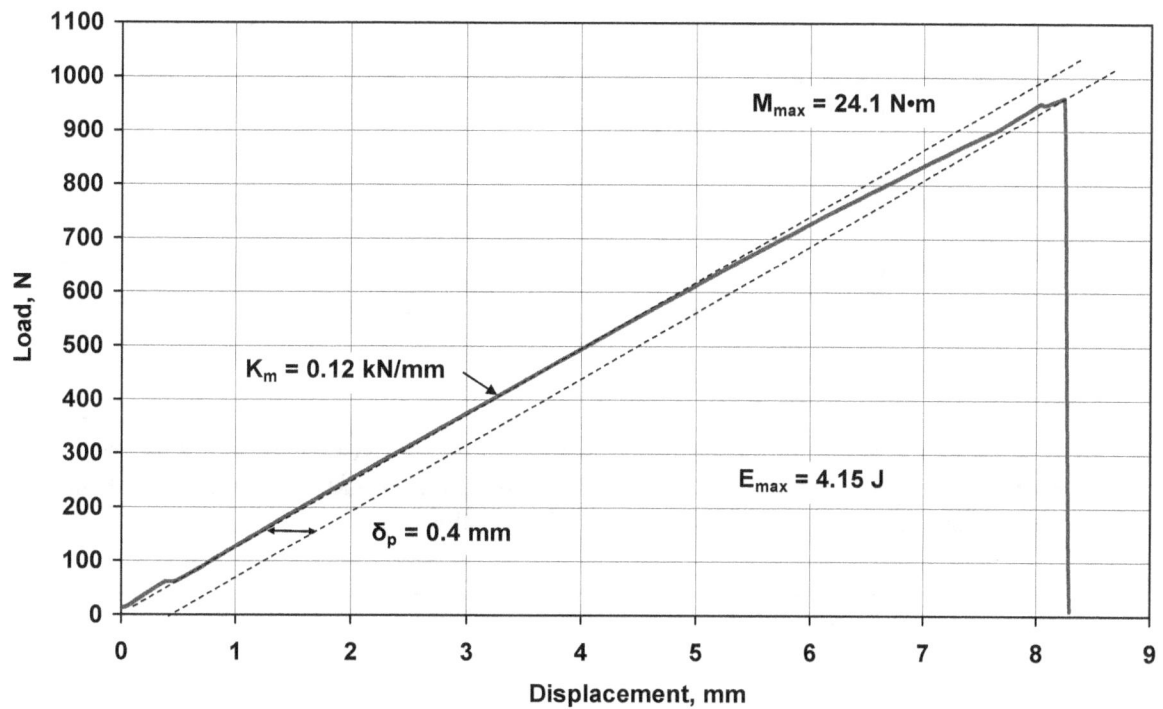

(a) Load-displacement curve

(b) Post-test sample

Figure 9. Results of 4-PBT conducted at 135°C and 1 mm/s with as-fabricated ZIRLO cladding oxidized (one-sided) at 1200°C to 17% CP-ECR and quenched at 800°C (Test OCZL#24): (a) load-displacement curve showing expected low ductility (0.4 mm offset displacement) and (b) post-bend-test appearance of central region of sample showing severing at two locations on either side of sample mid-span marked in ink on this specimen. The severed cross section to the right in Fig. 9b appears to be the primary failure location. The severing of the cross section to the left may have occurred during impact of the sample with the bottom of the test fixture.

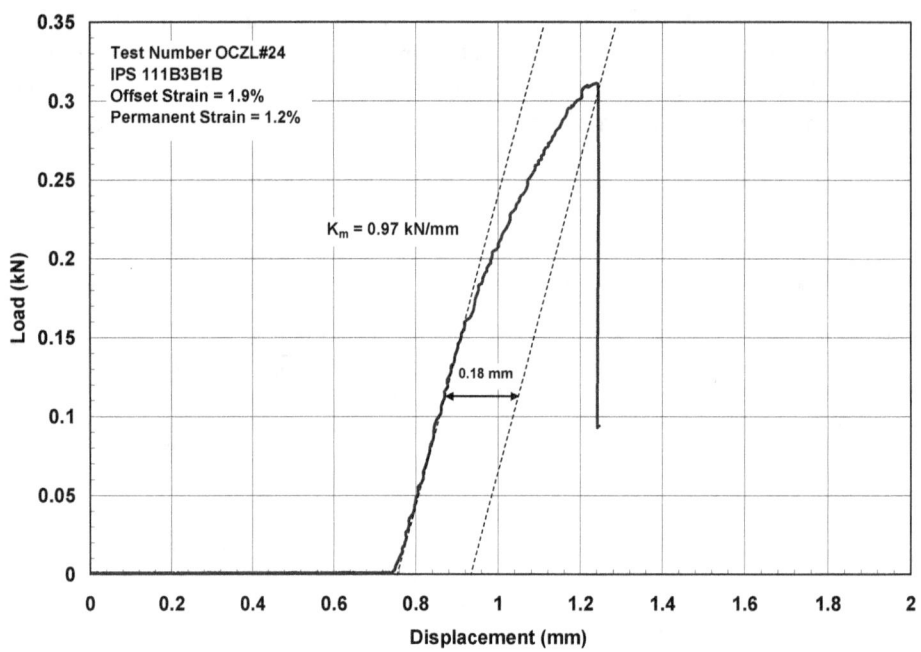

Figure 10. Load-displacement curve for Ring 1B sectioned from post-LOCA-bend sample OCZL#24 oxidized (one-sided) at 1200°C to 17% CP-ECR. Sample had one tight crack at the 6 o'clock position and 1.2% permanent strain (ductile).

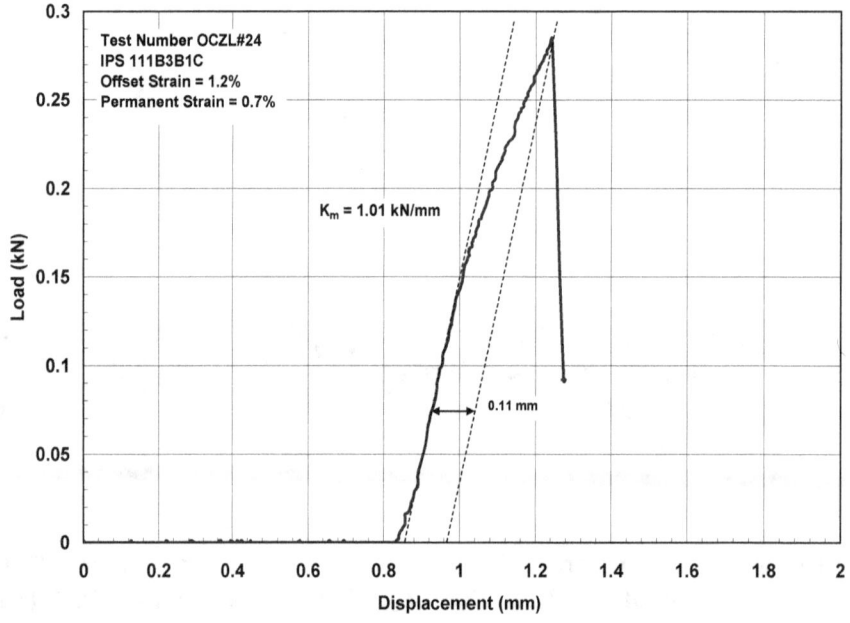

Figure 11. Load-displacement curve for Ring 1C sectioned from post-LOCA-bend sample OCZL#24 oxidized (one-sided) at 1200°C to 17% CP-ECR. Sample had one tight crack at the 12 o'clock position and 0.7% permanent strain (brittle).

16

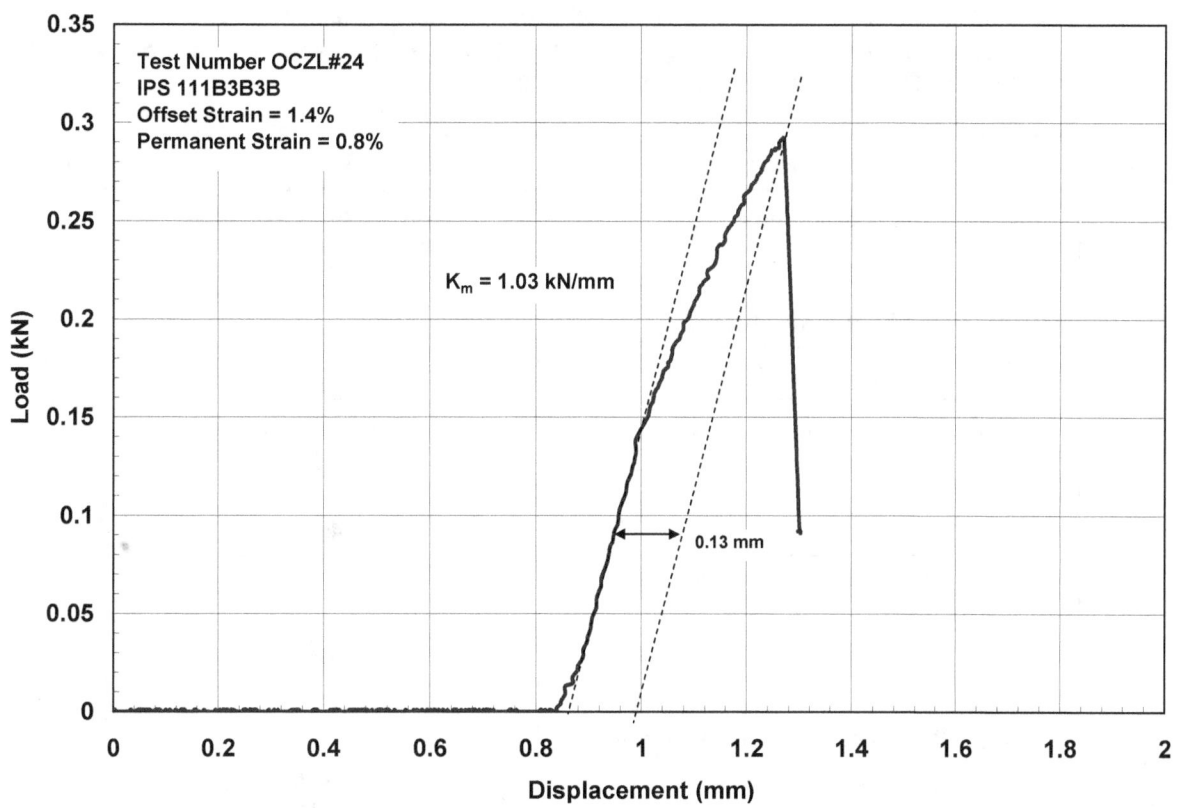

Figure 12. Load-displacement curve for Ring 3B sectioned from post-LOCA-bend sample OCZL#24 oxidized (one-sided) at 1200°C to 17% CP-ECR. Sample had one tight crack at the 6 o'clock position and 0.8% permanent strain (brittle).

The measured stiffness shown in Fig. 9a is 122 N/mm. This is the same stiffness as was calculated for non-oxidized 17×17 Zry-4 and measured for non-oxidized 17×17 M5. The oxidized ZIRLO described in Table 2 has a calculated bending stiffness of 156 N/mm assuming MATPRO [3] elastic constants for the oxide (14.8×10^4 MPa) and the metal (8.65×10^4 MPa) at 135°C. The measured stiffness is about 20% lower than the calculated stiffness. Some of this difference may be due to decreasing oxide-layer thickness from about 50 mm from the mid-span to the support locations.

It is interesting that the compressed rings developed a single crack at the 6-o'clock (2 rings) or 12-o'clock (1 ring) position. The crack initiated at the prior-beta layer inner surface and propagated through the alpha and oxide layers. Although CEA has conducted many ring compression tests using one-sided-oxidized samples, it has never been clear at which location the crack initiates because CEA does not stop tests at the first significant load drop. It has been assumed that the first crack initiates at the outer surface of the 3 or 9 o'clock position. This is clearly not the case for rings approaching the embrittlement oxidation level.

The 4-PBT requires longer samples than the 3-PBT, and this makes it more difficult to pre-hydride and oxidize the 4-PBT samples uniformly in a LOCA-type transient. In comparing the ANL 4-PBT sample to the CEA 3-PBT for non-ballooned specimens, the only advantage of the ANL 4-PBT is low sample stiffness, which allows for accurate determination of failure energy.

Table 2. Results of characterization and ring-compression testing (at 135°C and 2 mm/minute displacement rate) for the post-bend OCZL#24 sample that was oxidized (one-sided) at 1200°C to 17% CP-ECR and quenched at 800°C.

Parameters	Value
Oxidized Cladding OD, mm	9.73±0.02
OD Oxide Layer Thickness, µm	103±9
Average Metal Wall Thickness, mm	0.50
Average Metal ID, mm	8.52
RCT Offset Displacement, mm	0.18 0.11 <u>0.13</u> 0.14±0.04
RCT Offset Strain, %	1.9 1.2 <u>1.4</u> 1.5±0.4
RCT Permanent Displacement, mm	0.11 0.07 <u>0.08</u> 0.09±0.02
RCT Permanent Strain, %	1.2 0.7 <u>0.8</u> 0.9±0.3

2.4 Axial tensile Tests

Axial tensile tests have not been used to determine ductile-to-brittle transitions. Special grips would need to be designed to avoid failure at the grip locations. Also, machining is required to define a gauge section over which stress and strain can be determined based on load-displacement data. Metal gauge sections can be machined precisely by an electro-discharge machine (EDM) after outer-surface-oxide removal. This would be very difficult to do for post-LOCA samples because the oxygen-stabilized alpha layer is more brittle than the oxide layer. Oxide layer removal by mechanical means would result in damage to this alpha layer.

Axial tensile samples have a much higher stiffness ($K = EA/L$) than ring-compression and axial-bend samples. "A" is the gauge cross-sectional area and L is the gauge length. For 17×17 cladding tested at 135°C, sample stiffness values range from 27 kN/mm (51-mm-long ASTM sample with no gauge) to 9.9 kN/mm (ANL double gauge, 25.4-mm-long, 2.5-mm-width per gauge [7]). The long load train acts like a spring in series with the sample. High sample stiffness results in high load-train loads, small elastic strains along the load train, and load-train displacements larger than sample elastic displacements. The load cell accurately measures the sample load, but the measured displacement includes both load-train and sample displacements. Let $K_g = E_g A_g / L_g$ be the stiffness of the sample gauge section and K_m be the effective machine stiffness (includes all loaded components outside of gauge region). The total displacement measured by the Instron actuator is the sum of the machine displacement ($\delta_m = P/K_m$) and the sample gauge displacement ($\delta_g = P/K_g$).

$$\delta = \delta_m + \delta_g = P/K_m + P/K_g = (1/K_m + 1/K_g)\ P \tag{3}$$

Thus, the combined stiffness (K) can be found by solving Eq. 3 to get:

$$K = (K_g/K_m + 1)^{-1}\ K_g \tag{4}$$

Figure 13 shows (a) the load-displacement data and (b) the corresponding engineering stress-strain response for an as-fabricated HBR-type 15×15 Zry-4 ANL tensile sample tested at RT. The engineering strain is determined by the ratio of the measured total displacement (δ) and the gauge length ($L_g = 25.4$ mm). The stress is simply the load (P) divided by the cross-sectional area ($A_g = 3.8$ mm^2). For this sample, K_g is calculated to be 13.8 kN/mm. As shown in Fig. 13a, the load-displacement curve has a linearized loading slope of K = 4.65 kN/mm. Solving Eq. 4 for K_m gives, $K_m = 6.9$ kN/mm. Thus, for this particular tensile testing, the machine stiffness (includes grips and non-gauge part of sample) is about one-half of the sample gauge stiffness. The effect of these two "springs" in series gives a response stiffness about one-third of the sample gauge stiffness. This is easily seen in the Fig. 13b stress-strain plot. The slope of the elastic stress-strain curve is expected to be Young's modulus (E = 92,400 MPa), but the linearized slope of the stress-strain curve is only a third of that value. However, Fig. 13b also shows that the offset strain method can be used to determine the yield strength (≈600 MPa) at 0.2% offset strain. In principle, this methodology could be used to determine ductile-to-brittle transition (e.g., offset strains < 0.2% imply brittle behavior), but tensile tests with gauged samples are not practical for post-LOCA embrittlement studies.

2.5   Ring-stretch Tests

Hoop tensile tests have been performed to determine plastic stress-strain relationships for as-fabricated, pre-hydrided, and irradiated cladding alloys. The hoop tensile test requires machined gauge sections and extensive finite-element analysis to determine the plastic stress-strain relationship in the hoop direction [7]. The length of the sample is small (≈3 mm with 1-mm gauge), but the stiffness is relatively high (1.6 kN/mm for 17×17 Zry-4 at 135°C). Radial loading is induced by pulling on a pair of D-shaped mandrels or on a pair of modified D-shaped mandrels with a dog-bone insert to minimize bending. This test is not appropriate as a ductility screening test for post-LOCA samples because of pre-test machining requirements and difficulty in determining low-ductility values.

The Penn State University (PSU) plane-strain test has been used as a ductility test conducted with samples subjected to a biaxial stress state [7]. Sample preparation requires oxide removal, machined notches at the axial ends of the sample, and micro-indents on the cladding metal outer surface. Samples are longer than hoop-tensile samples (≈13 mm with ≈7 mm between notches). The stiffness of these samples is high (>10 kN/mm). Radial loading is induced by pulling on a pair of D-shaped mandrels. Plastic displacement is determined from permanent change in length between micro-indents. This test is not appropriate as a ductility screening test for oxidized post-LOCA samples for the same reasons that the uniaxial ring-stretch samples are not appropriate.

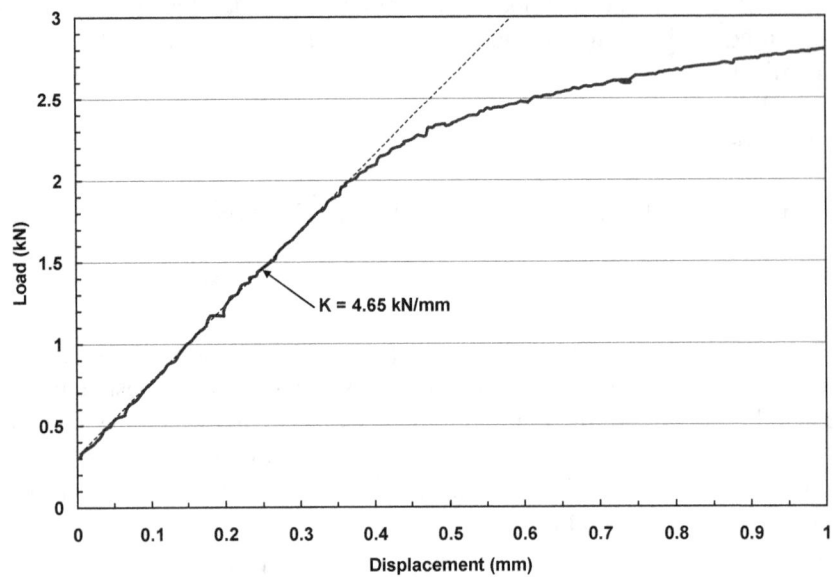

(a) Load-displacement curve

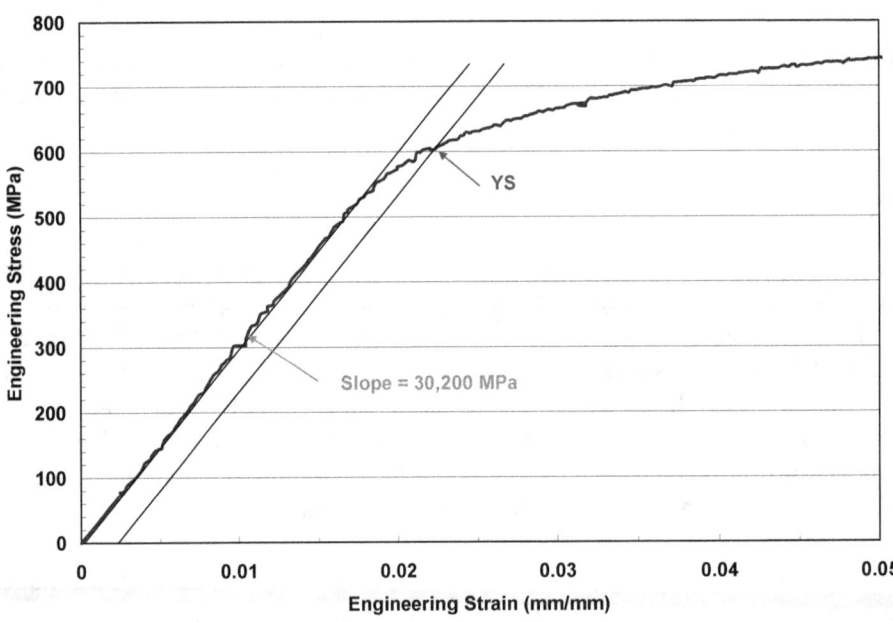

(b) Stress-strain curve

Figure 13. Load-displacement curve (a) and engineering stress-strain response (b) for 15×15 Zry-4 sample subjected to axial tensile loading at room temperature. The slopes of both curves are about one-third of the expected slopes based on the sample gauge length, cross-sectional area, and Young's modulus.

Another way of performing ring-stretch tests is the so-called expansion-due-to-compression (EDC) test method, for which pre-test machining of gauge sections is not needed [8,9]. Radial loading is induced by axial compression of a plug inserted into the sample and extending about 0.5 mm above the sample. Load is applied to the plug by the top ram or top-bottom rams. The hoop tensile stress induced in the sample is much more uniform than what can be achieved in the hoop-tension ring-stretch test. ORNL has used EDC to measure the RT hoop tensile properties of non-irradiated and irradiated Zry-4 [8]. Studsvik has used the same technique to measure the hoop tensile properties of irradiated Zry-2 in the temperature range of 25-340°C [9]. Typical cladding lengths for these tests are 7 mm (ORNL) and 20 mm (Studsvik). Stiffness values are >10 kN/mm for these samples. The expansion vs. time of the sample outer surface is measured directly by a pair of laser probes. The plug (polyurethane) is chosen to have a low hardness value. Expansion of the sample is limited by the plug displacement (≈1 mm for ORNL and 2.9 mm for Studsvik). Compression of the plug beyond the top edge of the sample may cause sample "barreling" and axial stresses (ORNL). Cladding diameter change is limited to ≈4% strain (ORNL) and ≈20% (Studsvik). The measured ram load and ram displacement represent the response of the ram-plug-sample system. Load-displacement curves for the ram cannot be used directly to determine sample offset displacements and strains. Determination of the hoop stress in the cladding as a function of the measured diameter or change in diameter is non-trivial. In addition to uncertainties in the elastic-plastic behavior of the plug, friction forces must be accounted for. Data plots are usually in the form of measured cladding radial displacement vs. ram displacement. A factor is derived by independent analyses or experiments to relate the measured ram load to the cladding hoop stress. ORNL determined this factor by using data from axial tensile tests for non-irradiated Zry-4 and assuming isotropy in the hoop and axial directions: $\sigma_\theta = 0.53$ $P/(h\ L)$. The same 0.53 factor was used for irradiated cladding. Although the EDC test has certain advantages relative to the other ring-stretch tests, it has not been used to measure the ductility of high-temperature, steam-oxidized cladding samples. This would require considerable development work and careful error analysis. It is anticipated that the relative error would increase as the cladding ductility decreased.

2.6 Comparison of Test Methods for Non-deformed Cladding

Table 3 shows a summary comparison of test methods for ductility and embrittlement determination of non-deformed oxidized cladding samples. Tests that require samples with end grips (e.g., axial tension) and/or machined gauge sections (e.g., axial and hoop tension) are not practical for LOCA-oxidized samples. Samples with high stiffness values relative to machine stiffness values are not desirable if one wants to compute failure energy, as well as ductility, from load-displacement curves. Tests that give load-displacement curves with a loading stiffness equal to the unloading stiffness are better if the offset displacement method is used to determine ductility. Also, tests with smaller samples are better for irradiated cladding. Taking all of these factors into account, the 3-PBT is the best choice for post-LOCA embrittlement determination of as-fabricated and pre-hydrided samples whereas the ring-compression test is the best choice for irradiated samples. However, it is sometimes desirable to have a single test method for as-fabricated, pre-hydrided, and irradiated cladding to eliminate systematic errors when comparing the behavior of irradiated and non-irradiated cladding. For such a comprehensive study, the RCT is preferred.

For the 3-PBT, the CEA span length of 80 mm has been chosen for calculations and comparison purposes. CEA 3-PBT results suggest that this span is adequate for determining offset displacement (δ_p) and embrittlement threshold, along with the criterion that $\delta_p \geq 0.1$ mm implies ductility. However, if the CEA 3-PBT were to be used to determine failure energy, the span length could be increased from 80 mm to 110 mm (reduces the stiffness from 1300 N/mm to 500 N/mm, as stiffness varies inversely with the cube of the span length) and a relatively stiff machine ($K_m > 5000$ N/mm) could be used. This would result in better agreement between the calculated and measured (within 10%) loading stiffness and failure energy values. Although not currently used in ductility screening tests, the failure energy (measure of toughness) is a useful parameter to determine. The same machine could be used for RCT samples, which would result in better agreement between the measured and calculated slopes of the loading curve. At 135°C, the predicted loading slope for as-fabricated 17×17 Zry-4 is 1000 N/mm. The ANL-measured RCT loading slope for CEA samples (600-wppm-H 17×17 Zry-4) was 820 N/mm following oxidation and quench. CEA could conduct benchmark RCTs and 3-PBTs with as-fabricated 17×17 Zry-4 to determine if there is a significant difference between calculated and measured loading slopes. The results of such simple tests would provide useful information regarding the stiffness of the machine and fixtures used to conduct these tests.

Table 3. Comparison of test methods for determination of ductile-to-brittle-transition oxidation level of post-LOCA samples at 135°C. Sample stiffness values are given qualitatively relative to machine stiffness values. 3-PBT and 4-PBT represent 3- and 4-point bend tests, respectively. EDC is expansion due to compression. PSU is Penn State University. R&D = research and development.

Test Type	Sample Length, mm	Grips Required	Gauge Sections Required	Sample Stiffness	Determination of Offset Displacement or Strain From Load-Displacement Curve	Ranking for Non-irradiated Cladding	Ranking for Irradiated Cladding
ANL-CEA Ring Comp. Test	8-10	No	No	Low	Good	High	Very High
CEA 3-PBT	90	No	No	Low	Very Good	Very High	High
ANL 4-PBT	300	No	No	Very Low	Very Good	Low	Low
ANL Axial Tension Test	76	Yes	Yes	High	OK for ≥0.2% strain	Low	Low
ANL Hoop Tension Test	3	No	Yes	Moderate	Difficult	Low	Low
PSU Plane Strain Test	13	No	Yes	High	Difficult	Low	Low
EDC Test ORNL Studsvik	7 20	No No	No No	High High	Difficult Difficult	Needs R&D Needs R&D	Needs R&D Needs R&D

3   TEST METHODS FOR BALLOONED AND RUPTURED CLADDING

Long, pressurized cladding samples subjected to a LOCA transient can balloon and rupture. Such samples will contain local areas that exhibit considerable axial variation in diameter, wall thickness, hydrogen content, and oxidation level. A substantial circumferential variation in wall thickness and oxidation level will also occur in cross sections containing a rupture opening (see Refs. 1, 10-12). Figure 14 shows the variation in wall thickness at the mid-span rupture cross section for the ANL Test OCZL#18 sample which had 43% maximum circumferential strain. The maximum oxidation level (12% CP-ECR based on average wall thickness prior to oxidation) occurred in this cross section for which the average metal wall thickness prior to oxidation was 0.40 mm. Consistent with current licensing guidelines in 10 CFR 50.46(b), all CP-ECR values reported in this section for the rupture-region cross section are based on average wall thickness prior to oxidation.

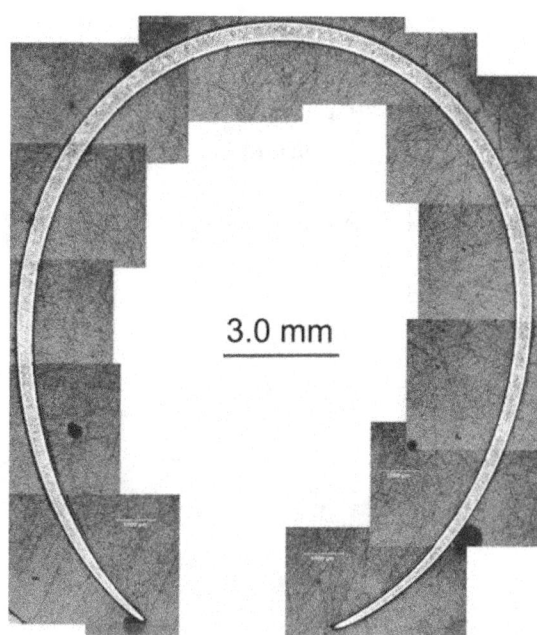

Figure 14. Low magnification image of cross section through rupture mid-span for Test OCZL#18 sample with 43% mid-wall circumferential strain oxidized to 12% CP-ECR based on a pre-oxidation average wall thickness of 0.40 mm.

Figure 15 shows the axial variation for the OCZL#18 sample in (a) cladding diametral strain at two orientations and (b) cladding hydrogen content. Figure 16 shows axial profiles for (a) diametral strains and (b) hydrogen and oxygen content for the OCZL#19 test sample with 24% maximum circumferential strain and 17% oxidation level. While the maximum oxidation level occurred at the rupture-opening mid-span, the maximum hydrogen pickup occurred outside the rupture opening close to the necks of the balloon. These tests were conducted using as-fabricated 17×17 ZIRLO cladding filled with zirconia pellets, pressurized to 1200 psig (OCZL#18) or 600 psig (OCZL#19) at 300°C, ramped at 5°C/s to 1200°C in steam, held at 1200°C, cooled at 3°C/s to 800°C, and rapidly cooled via water quench from 800°C to 100°C.

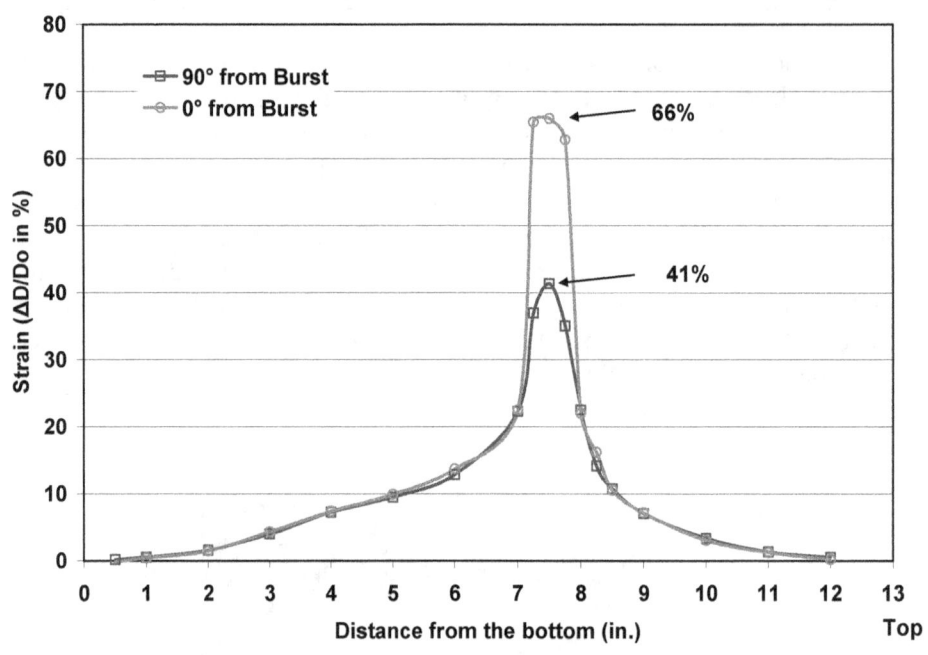

(a) Axial profile of diametral strains

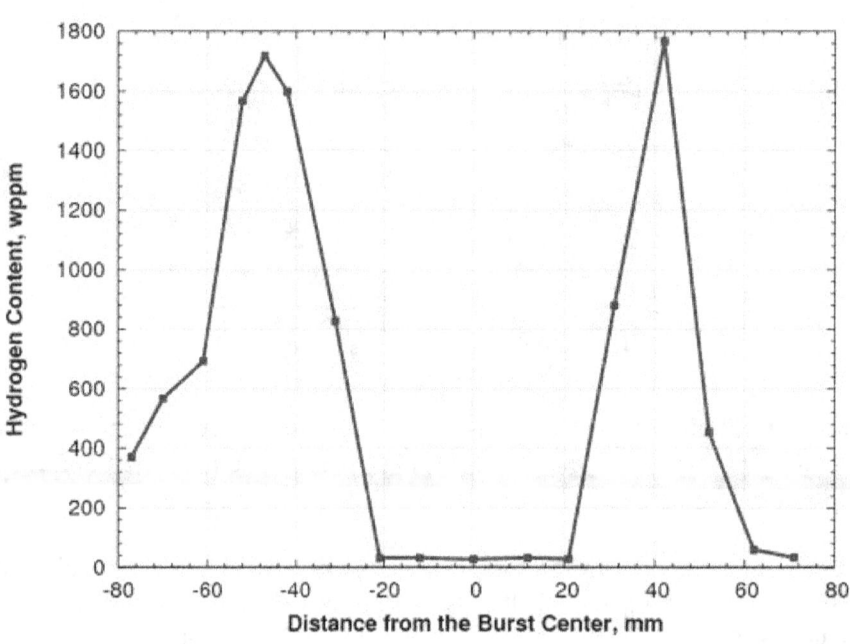

(b) Axial profile of hydrogen content

Figure 15. Axial profiles of (a) cladding diametral strains and (b) hydrogen content for the OCZL#18 test sample with 43% maximum circumferential strain oxidized to 12% CP-ECR at T ≤ 1200°C, and quenched at 800°C. D_o was 9.50 mm.

26

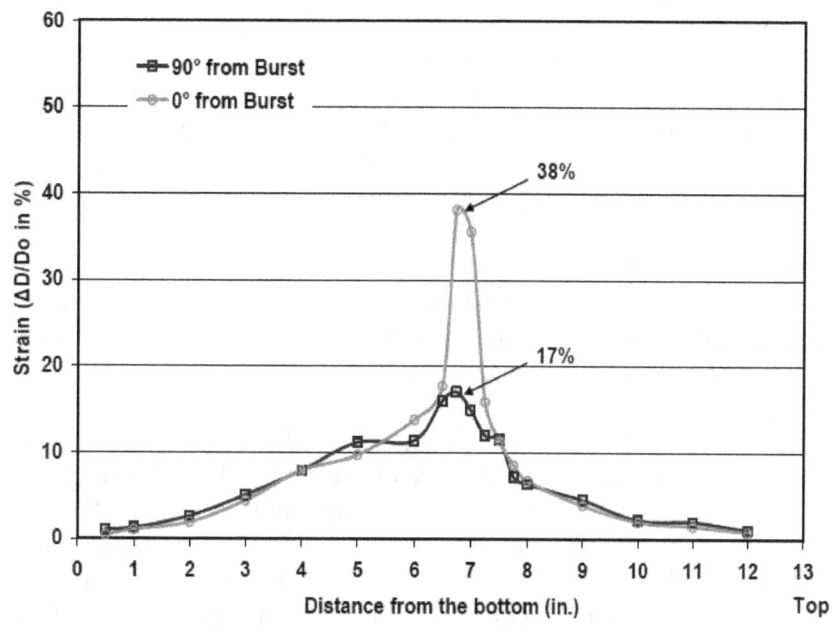

(a) Axial profile of diametral strains

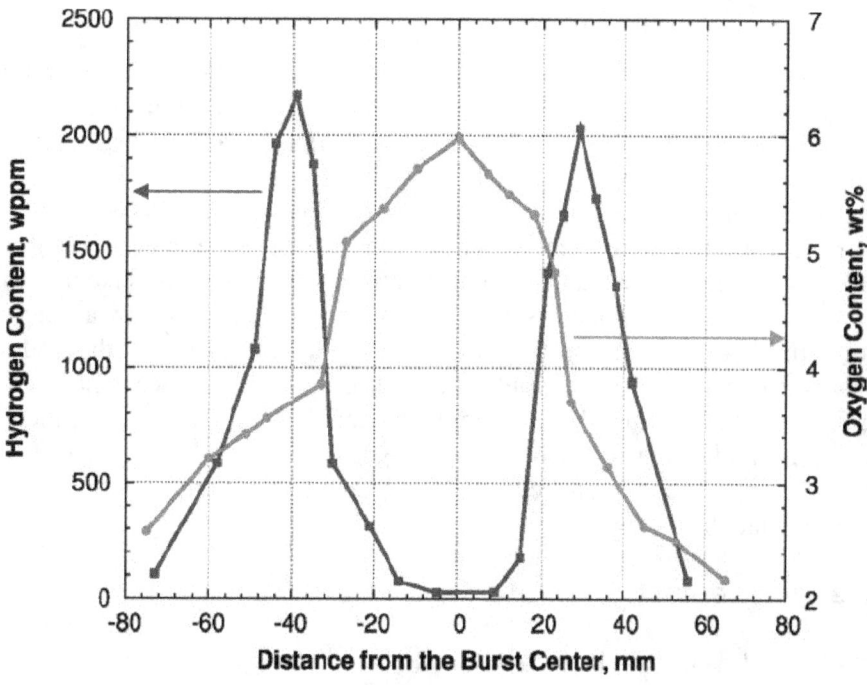

(b) Axial profile of hydrogen and oxygen content

Figure 16. Axial profiles of (a) cladding diametral strains and (b) hydrogen and oxygen content for the OCZL#19 test sample with 24% maximum circumferential strain oxidized to 17% CP-ECR at T ≤ 1200°C and quenched at 800°C. D_o was 9.50 mm.

Ring-compression tests with samples sectioned from the balloon region are not appropriate tests because of the steep variation in outer diameter. Loading would be highly localized at the point of maximum outer diameter, and it would spread axially as the cladding deformed or cracked. This would result in an increase in the load-displacement slope prior to plastic deformation and/or cracking. Also, results would be highly dependent on orientation of the loading location relative to the rupture opening. The determination of "ductility" prior to cracking along the length of the sample would be highly uncertain. Therefore, axial tensile and bend tests are more appropriate for studying ballooned specimens, and such testing with ballooned specimens is relatively recent.

To produce ballooned specimens for axial tensile and bend tests, ANL and the Japanese Atomic Energy Agency (JAEA) [13-14] at their Tokai research center have performed LOCA integral experiments with lengths of as-fabricated and pre-hydrided cladding alloys. In addition, JAEA has conducted experiments with defueled cladding sectioned from irradiated rods [15-16].

JAEA integral experiments differ from ANL's in terms of sample length, internal pressure and volume, heating rate, hold temperature, cooling rate and quench temperature. Experimental parameters are compared in Table 4. (Not included in Table 4 are JAEA high-burnup Zry-2, ZIRLO, MDA and NDA samples [16].) However, the most significant difference between ANL and JAEA testing has to do with mechanical test methods. ANL has performed post-quench 4-PBTs at 135°C, while JAEA has performed full and partial axial-restraint tests during cooling from the hold temperature to <100°C. These mechanical test methods are reviewed in Sections 3.2 and 3.3.

3.1 Axial Tensile Tests

Traditional axial tensile tests could be performed on post-LOCA integral samples, and pulling the sample to failure would give an accurate measure of failure load, as every cross section is subjected to the same tensile load. The samples are long, have welded end caps for gripping, and are highly unlikely to fail at the grip locations because these are well outside the uniform temperature zone. The oxidation level and hydrogen content at these end locations are both low relative to the middle of the sample. However, during the pressurized phase of the transient, the sample would experience some bending (see Fig. 17 for ANL 17×17 ZIRLO sample). Subjecting this sample to a displacement-controlled axial tension test would result in a low initial loading slope to straighten out the sample followed by a steeper slope due to combined machine stiffness and sample stiffness (see Eqs. 3 and 4). However, if the load-displacement curve exhibited any offset displacement, it would not be clear if plastic displacement had occurred within the ruptured region of the balloon, just above and/or below this region where the oxidation temperature decreases, or outside the middle region where the cladding is annealed and lightly oxidized with negligible hydrogen pickup. Therefore, the only meaningful data that could be derived from traditional axial tensile tests are failure loads.

3.2 Axial Restraint Tests

The JAEA LOCA integral samples were not tested in the traditional way, but were restrained from axial contraction during quench in the integral experiment. The purpose of these tests was to determine: the fracture/no-fracture boundary as a function of hydrogen content and oxidation level and the fracture load for tests in which fracture occurred. Both full- and partial-axial-restraint tests have been conducted during cooling from the oxidation temperature to <100°C. Figure 18a shows the gripping mechanism used to restrain the cladding. Figure 18b shows typical axial load vs. time results for fully restrained and

28

partially restrained cases that were limited to maximum axial loads of 735 N, 540 N (reference JAEA case), and 390 N.

Table 4. **Comparison of ANL and JAEA [13-16] LOCA integral sample and experimental parameters. AF is as-fabricated, PH is pre-hydrided, and Irr is irradiated.**

Parameters	ANL		JAEA		
	AF	PH	AF	PH	Irr
17×17 Cladding	ZIRLO	ZIRLO	Zry-4	Zry-4	Zry-4
Cladding OD, mm	9.50	9.50	9.42	9.42	---
Wall Thick., mm	0.57	0.57	0.51	0.51	---
Sample Length, mm (minus end caps)	295	295	570	570	180
Hydrogen Content, wppm	≈10	200-600	≈10	100-1400	150±40
Pellets	zirconia	zirconia	alumina	alumina	alumina
Pellet Stack Length, mm	280	280	550	550	170
Gas Volume, cm^3	10	10	4.4	4.4	2.4
Internal Pressure, MPa (Gauge)	4.14 8.28 @ 300°C	4.14 8.28 @ 300°C	5 @ RT	5 @ RT	5 @ RT
Heating Rate, °C/s	5	5	10	10	10
Rupture T, °C	843±6 750±7	742 700±30	---	656-880	786±25
Rupture Strain, %	22±3 46±4	58 60±10	35 (avg.)	11- 40	16±8
Hold Temp., °C	1200	1200	947-1257	947-1257	1030-1178
Cooling Rate to Quench Temp., °C/s	3	3	20 to 900°C 5 to 700°C	20 to 900°C 5 to 700°C	20 to 900°C 5 to 700°C
Quench Temp., °C	800	800	700	700	700

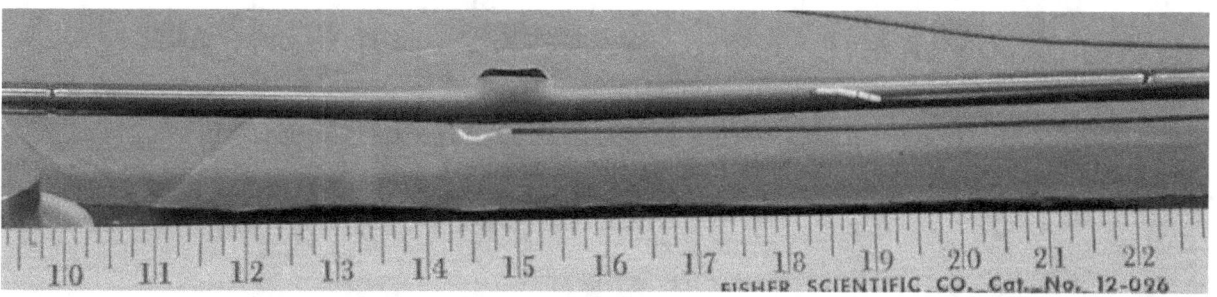

Figure 17. Shape of the ANL OCZL#29 post-quench sample with maximum circumferential strain of 49% and maximum CP-ECR of 17%.

29

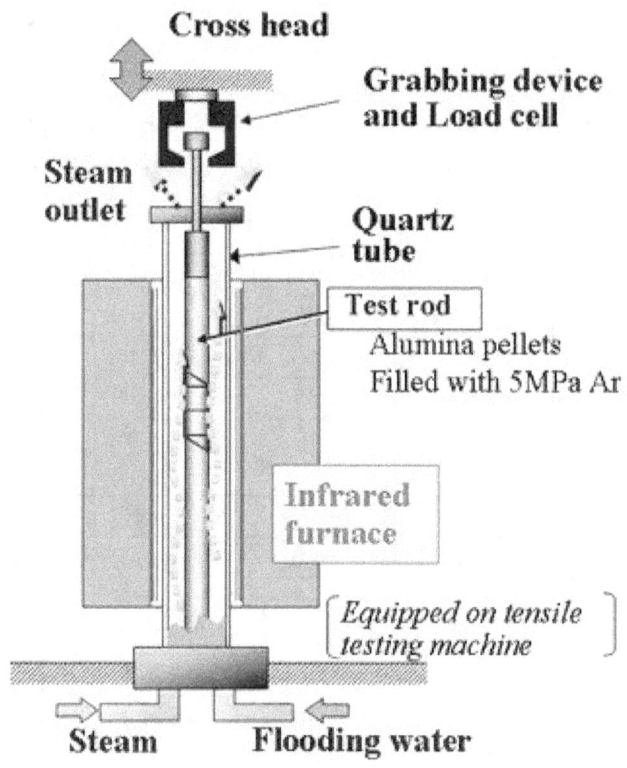

(a) JAEA LOCA apparatus

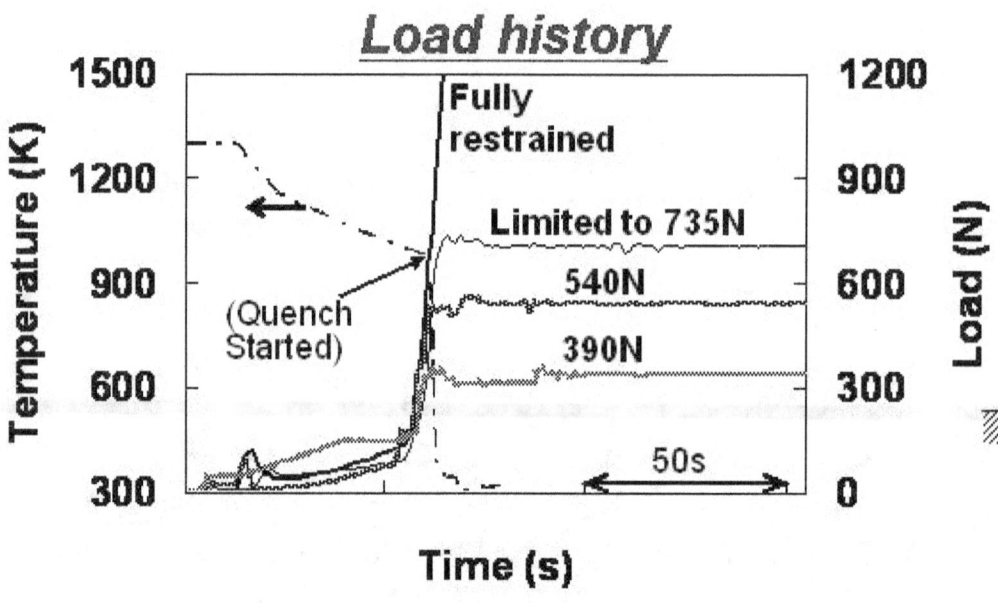

(b) Axial tensile loads vs. time for several constraint conditions

Figure 18. JAEA LOCA apparatus (a) showing the gripping device used to restrain sample during cooling and (b) load vs. time curves for fully restrained samples and partially restrained (maximum loads of 735 N, 540 N, and 390 N) samples [17].

The measured load is relatively low in these tests during cooling from the oxidation hold temperature to the 700°C quench temperature due to the transition from a bent sample to a straight sample and due to the low yield strength and high plasticity of the cladding metal at elevated temperatures. However, the load builds up rapidly during quench cooling because the metal yield strength increases with decreasing temperature and the metal does not have time to flow plastically to relax the thermal stresses.

For the fully restrained cases [13-14], the maximum load was 1200 to 2400 N for samples that survived quench without failure. For samples that failed during quench, the failure loads were 200 to 1700 N. Most of the failures occurred in the rupture node, which is not expected to pick up much (<30 wppm) hydrogen during the oxidation phase. These are important results, however, because regardless of the degree of restraint, the maximum axial load due to differential contraction was found to be ≤2400 N.

JAEA conducted 118 tests with pre-hydrided cladding, which included variation of the hydrogen content, oxidation hold temperature, oxidation hold time, and degree of axial constraint. Forty-seven samples failed during quench with 42 of the failures occurring in the rupture node.

JAEA also conducted 6 tests using irradiated Zry-4 cladding from PWR fuel rods with ≤44 GWd/MTU burnup, ≤25-μm corrosion layer, and estimated hydrogen contents in the range of 150±40 wppm. Based on several analyses of the strength of grid spacers, JAEA argued that axial restraint loads had to be < 1000 N. They adopted 540 N for their partially restrained tests using irradiated cladding samples.

For irradiated cladding, the sample length was reduced from 580 mm to 190 mm. Notice from Table 4 that the RT gas volume was reduced from 4.4 cm^3 to 2.4 cm^3. Rupture strains were smaller (16±8%) than measured for longer, pre-hydrided test samples of comparable hydrogen content. The smaller gas volume and the shorter (≈40 mm) uniform temperature zone may have contributed to the lower rupture strains. Oxidation temperatures were 1171±9°C for five of the tests and 1030°C for one of the tests. Partially restrained tensile tests were conducted during cooling with the maximum axial load set at 540 N. Two of the samples oxidized at about 1170°C failed at axial loads of 498 N (170-wppm H and 23% CP-ECR) and 385 N (120 wppm hydrogen and 20% CP-ECR).

For the sample that fractured at 498 N, post-test imaging (see Fig. 19) indicated that failure occurred near the edge of the rupture opening. JAEA interpreted this as a rupture-node failure. However, the results are open to interpretation. In particular, the high-hydrogen concentrations near the fracture are indicative of regions outside the rupture opening, which have high hydrogen pickup from inner-surface oxidation. Consistent with JAEA interpretation, the crack may have initiated at the edge of the rupture opening, but it clearly propagated into the high hydrogen zone outside the rupture opening.

For non-irradiated cladding, the hydrogen peaks were 30 to 50 mm from the rupture mid-span. Based on the results shown in Fig. 19, one hydrogen peak appears to occur at <10 mm from the rupture mid-span for irradiated Zry-4. It is not clear whether this was due to a difference between irradiated and pre-hydrided cladding or differences in sample length (190 mm vs. 580 mm), uniform temperature zone (40 mm vs. 100 mm), and/or rupture strain (14% vs. 20-40%).

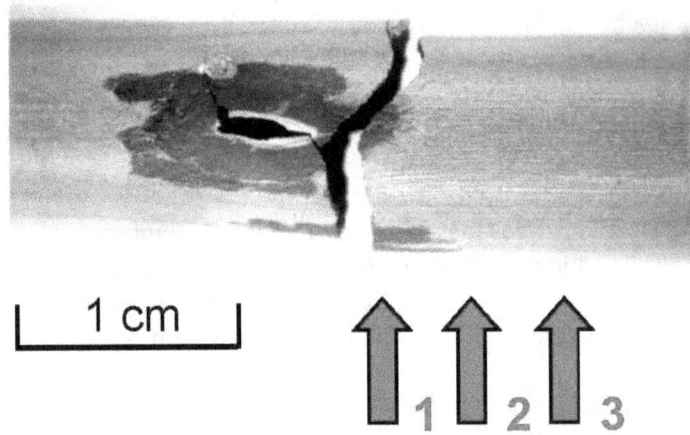

Hydrogen concentration
1. 1450 ppm
2. 1850 ppm
3. 840 ppm

Figure 19. Appearance [17] of JAEA A 3-1 sample with about 170-wppm pre-test hydrogen, oxidized at 1176°C to 29.3% BJ-ECR (23% CP-ECR), and partially constrained during cooling to a maximum axial tensile load of 540 N. Measured failure load was 498 N.

3.3 Axial Bend Tests

There are three locations (see Fig. 16b) in as-fabricated LOCA integral samples that are vulnerable to failure (i.e., severing of the cross-section) in response to axial tensile stresses induced by axial bending: the rupture location with the thinnest average wall thickness and therefore the highest oxidation level; and two locations on either side of the rupture location with higher hydrogen content and intermediate oxidation levels. These two locations are between the edges of the rupture opening and the hydrogen peaks. Hydrogen enhancement of oxygen embrittlement is responsible for failure at the intermediate locations. For as-fabricated cladding, the hydrogen pickup is negligible (<30 wppm) in the cross sections containing the rupture opening.

As discussed in Section 2.2, the 3-PBT with the load applied to the thick back of the rupture-opening cross section would bias the failure to occur in the cross section under the load that contained the rupture opening. The 4-PBT does not bias the failure location because the bending moment is uniform over the span L_s (see Fig. 7). As such, ANL has used 4-PBTs to determine post-LOCA sample failure location, maximum bending moment (measure of strength), failure energy (measure of toughness), and offset displacement (measure of plastic deformation).

The bending moment M is determined from the applied lateral force as shown in Fig. 7. With analysis, the bending moment can also be related to an equivalent axial tensile load (see Section 3.4). The maximum energy E_{max} is determined by calculating the area under the load-displacement curve and dividing by 1000 mm/m to express E_{max} in joule energy units. For samples that fail during 4-PBTs, the failure energy is the maximum energy. The offset displacement is determined at the loading points (see

Figs. 7 and 8a). ANL 4-PBTs were conducted in the standard displacement-controlled mode at 2 mm/s (later reduced to 1 mm/s) and a maximum displacement at the loading points of 14 mm. In accordance with ASTM standards for fracture toughness and Charpy impact tests, the load was applied to cladding at 180° relative to the rupture opening in order to subject the flawed rupture tips to maximum tensile stresses. Two tests were conducted with the rupture tips subjected to maximum compressive stresses to determine the effects of sample orientation relative to the bending moment.

Table 4 summarizes the ANL test conditions for ballooning, rupture, oxidation, and quench of pressurized, as-fabricated 17×17 ZIRLO LOCA integral samples. Table 5 summarizes the 4-PBT results. With the exception of the OCZL#32 test sample, which was subjected to 4-PBT at RT, bending tests were conducted at 135°C. Table 5 contains three metrics for assessing cladding performance: maximum bending moment, failure or maximum energy, and offset displacement. Also included in Table 5 are the failure locations relative to the center of the rupture opening. LOCA integral samples with circumferential strains ≤32% all failed in a region where the cladding was fully brittle between the rupture tips and the hydrogen peaks. Six out of nine samples with ≥40% rupture strain failed in the rupture node location where some of the cladding had significant local ductility (e.g., thick back side of the balloon). Results for the three metrics are discussed in the following.

Maximum Bending Moment

Three ramp-to-rupture samples with rupture strains ranging from 21% to 69% were subjected to 4-PBTs to determine reference values for maximum bending moment and energy at ≈0% oxidation level (CP-ECR). After 14-mm displacement, the load-displacement curves were relatively flat, indicating that the end-of-test bending moments were close to their maximum values.

The maximum bending moment values for ZIRLO samples are plotted in Fig. 20 where it is seen that bending moment is a strong function of cladding oxidation that occurs after rupture.

Figure 20a shows results from 0% to 23% (maximum tested) CP-ECR. The data points for low rupture-strain samples that failed during 4-PBTs are within the scatter band of high rupture-strain samples. Figure 20b shows data for failed samples tested at 135°C to emphasize the data trend for ≥10% CP-ECR. For these samples, the best-fit linear correlation to the data is given by:

$$M_{max} = 13.92 - 1.073\,(CP\text{-}ECR - 10\%), \; N\bullet m \tag{5}$$

Equation 5 represents a reasonable fit to the data for CP-ECR values ≥10%. The upper oxidation limit for Eq. 5 is 19.2% at which the failure bending moment (4.1 N•m) is equal to the failure bending moment at about 23% CP-ECR. Thus, the oxidation limits for Eq. 5 are 10% to 19% CP-ECR. At the current licensing limit of 17%, Eq. 5 gives 6.4 N•m as the failure bending moment. Measured values in the range of 16-18% CP-ECR and 135°C test temperature were 4.7 to 8.3 N•m.

In order to achieve oxidation levels as low as about 11% CP-ECR (Tests OCZL#21 and #22) with a maximum oxidation temperature of 1200°C, it was necessary to increase the average pre-oxidation cladding wall thickness by lowering the rupture strain to about 20%. This change made the rupture opening cross section stronger than the higher hydrogen regions, in which severing then occurred. Similar results were obtained for all samples with ≤32% rupture strain.

Table 5. Summary of results for LOCA integral and post-LOCA bend tests with as-fabricated ZIRLO cladding. Reference LOCA test conditions were: 600- or 1200-psig fill pressure at 300°C, 5°C/s heating rate to 1200°C, 1200°C hold temperature, 3°C/s cooling rate to 800°C, and quench at 800°C. Reference conditions for 4-PBTs were 135°C test temperature and 2 mm/s displacement rate to 14-mm maximum displacement. The displacement rate was lowered to 1 mm/s after the OCZL#21 bend test.

Test ID OCZL#	Fill Pressure, psig	Rupture Strain, % (T_R, °C)	CP-ECR %	Quench at 800°C	Stress in Rupture Node	Failure Location	Maximum Bending Moment N•m	Maximum Energy J	Offset Displace. mm
8	600	21 (845±25)	0	No	Maximum tension	No cracking	20.9	>8.4	>7.7
9	400	33 (875±15)	0	No	Maximum tension	No cracking	20.6	>8.3	>7.7
10	1600	69 (715±10)	0	No	Maximum tension	No cracking	19.5	>7.7	>7.1
12	1000	32 (805±20)	14.3	No	Maximum compression	-40 mm +33 mm	10.5	0.78	0
13	1200	41 (741±15)	14.4	No	Maximum tension	Rupture opening	8.8	0.58	0
14	1200	47 (735±6)	18.2	Yes	Maximum tension	Rupture opening	5.7	0.24	0
15	1200	51 (755±23)	18.3	Yes	Maximum compression	Cracking; no failure	8.9	>2.3	>13
17	1200	49 (750±17)	13.1	Yes	Maximum tension	Rupture opening	8.4	0.71	>0.5
18	1200	43 (748±4)	12.3	Yes	Maximum tension	Rupture opening	13.5	1.29	0

Table 5. Summary of results for LOCA integral and post-LOCA bend tests with as-fabricated ZIRLO cladding. Reference LOCA test conditions were: 600- or 1200-psig fill pressure at 300°C, 5°C/s heating rate to 1200°C, 1200°C hold temperature, 3°C/s cooling rate to 800°C, and quench at 800°C. Reference conditions for 4-PBTs were 135°C test temperature and 2 mm/s displacement rate to 14-mm maximum displacement. The displacement rate was lowered to 1 mm/s after the OCZL#21 bend test. (Cont'd)

Test ID OCZL#	Fill Pressure, psig	Rupture Strain, % (T_R, °C)	CP-ECR %	Quench at 800°C	Stress in Rupture Node	Failure Location	Maximum Bending Moment N•m	Maximum Energy J	Offset Displace. mm
19	600	24 (840±12)	17.2	Yes	Maximum tension	+23 mm -23 mm	5.7	0.23	0
21	600	27 (850±10)	10.5	Yes	Maximum tension	+33 mm -29 mm	13.8	1.17	0
22[a]	600	22 (837±12)	11.5	Yes	Maximum tension	+25 mm -27 mm	11.1	0.83	0
25[a]	1200	42 (757±21)	16.1	Yes	Maximum tension	-26 mm +26 mm	8.3	0.50	0
29[a]	1200	49 (746±19)	17.4	Yes	Maximum tension	Rupture opening	4.7	0.40	>8.5
32[a,b]	1200	49 (748±8)	17.2	Yes	Maximum tension	Rupture opening	6.7	0.26	0
37[a]	1200	46 (755±30)	23.1	Yes	Maximum tension	-40 mm +35 mm	4.1	0.13	0
43[a]	1200	50 (738±16)	17.6	Yes	Maximum tension	+35 mm	6.1	0.24	0

[a]Displacement rate lowered to 1 mm/s to better match maximum elastic strain rate for ring-compression tests.
[b]4-PBT conducted at 30°C.

Based on one RT 4-PBT (OCZL#32 at 17% CP-ECR), the maximum bending moment appears to be relatively insensitive to test temperature. This result is quite different from RCT ductility results, which indicated that permanent and offset strains were strong functions of test temperature (135°C vs. RT).

The 4-PBT strength results appear to be a meaningful measure of cladding performance in that the strength exhibits an expected decrease with increasing oxidation level. However, application of these results to LOCA acceptance criteria is not straightforward. Bending moments and axial bending stresses during quench are not anticipated to be significantly large, especially compared to potential axial stresses from partial contraction restraint. However, JAEA results for intact, fully restrained cladding gave an upper bound on the maximum axial force (1200 to 2400 N) that could be generated during quench following oxidation at 1200°C. It is also shown in Section 3.4 that a failure bending moment of ≈6 N•m corresponds to an axial failure load of ≈2500 N. Thus, it appears that the current 17% oxidation limit may be sufficient to ensure that the balloon region of low-burnup cladding would remain intact following quench.

Failure Energy

The failure or maximum (when no failure occurs) energy is plotted vs. oxidation level in Fig. 21 where it is seen that the failure energy is also a strong function of cladding oxidation that occurs after rupture. Load-displacement results from ramp-to-rupture tests (0% CP-ECR) were used to determine upper-bound energies of ≈8 J for unfailed samples through 14-mm displacement. Although this value does not represent the maximum energy that such samples could accumulate prior to failure, it does represent the maximum that can be accumulated through 14-mm displacement. As can be seen in Fig. 21a, there is a sharp decrease in failure energy with an increase in oxidation level from 0 to 10%. The decrease in failure energy is more gradual for increasing oxidation levels from 10 to 19%. It appears to level off to about 0.13 J in the oxidation range of 19 to 23% CP-ECR. This can be better seen in Fig. 21b for which the data are limited to samples that failed during 4-PBT loading at 135°C.

Similar to the maximum bending moment results, there was no significant difference between failure energies for small (≤33% rupture strain) and large (≥40% rupture strain) balloons. At about 12% CP-ECR, the failure energies for 22% (OCZL#22) and 43% (OCZL#18) rupture-strain samples were 1.17 J and 1.29 J, respectively.

The best-fit linear correlation for failure energy vs. CP-ECR in the range of 10 to 19% oxidation level is given by:

$$E_{max} = 1.225 - 0.1236 \text{ (CP-ECR} - 10\%), \text{ J} \qquad (6)$$

The failure energy represented by Eq. 6 gives 0.36 J and 0.11 J at 17% CP-ECR and 19% CP-ECR, respectively. The measured value at 23% CP-ECR was 0.13 J. More data would be needed in the range of 18 to 23% CP-ECR to confirm this leveling off of failure energy. However, the results clearly show that the oxidation level should be limited to maintain a high level of toughness.

36

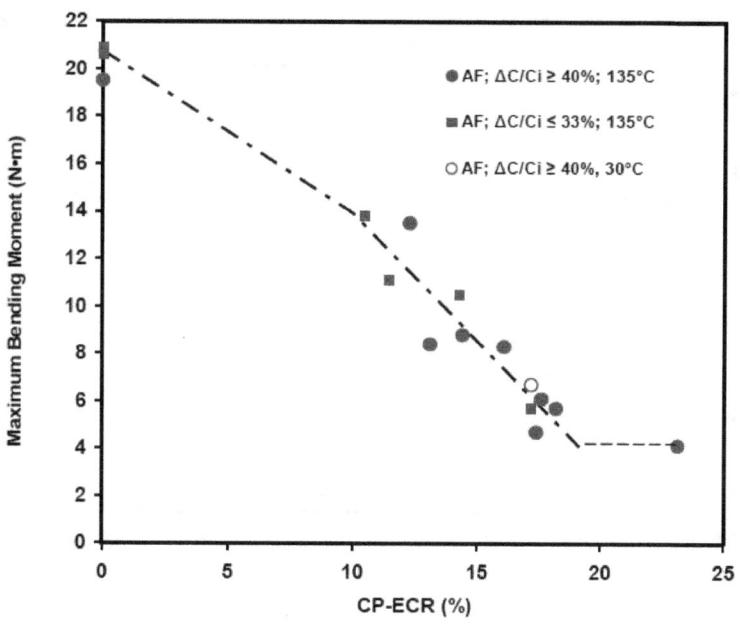

(a) Data for all samples subjected to 4-PBTs

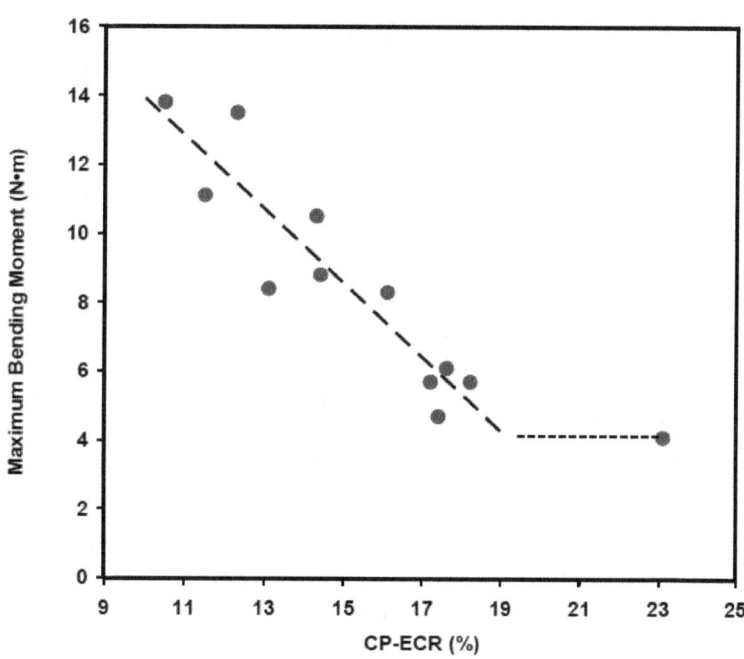

(b) Data for failed samples subjected to 4-PBTs at 135°C

Figure 20. Maximum bending moment as a function of maximum oxidation level (CP-ECR) for post-LOCA samples subjected to 4-PBTs with the rupture region in tension for all but one test. 4-PBTs were performed at 135°C and 2 or 1 mm/s to 14-mm maximum displacement. One bend test was conducted at 30°C.

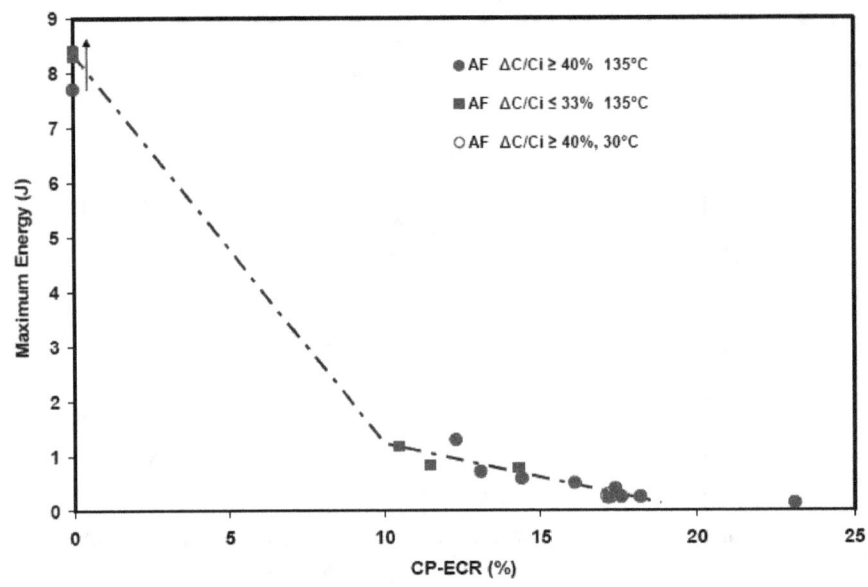

(a) Data for all samples subjected to 4-PBTs

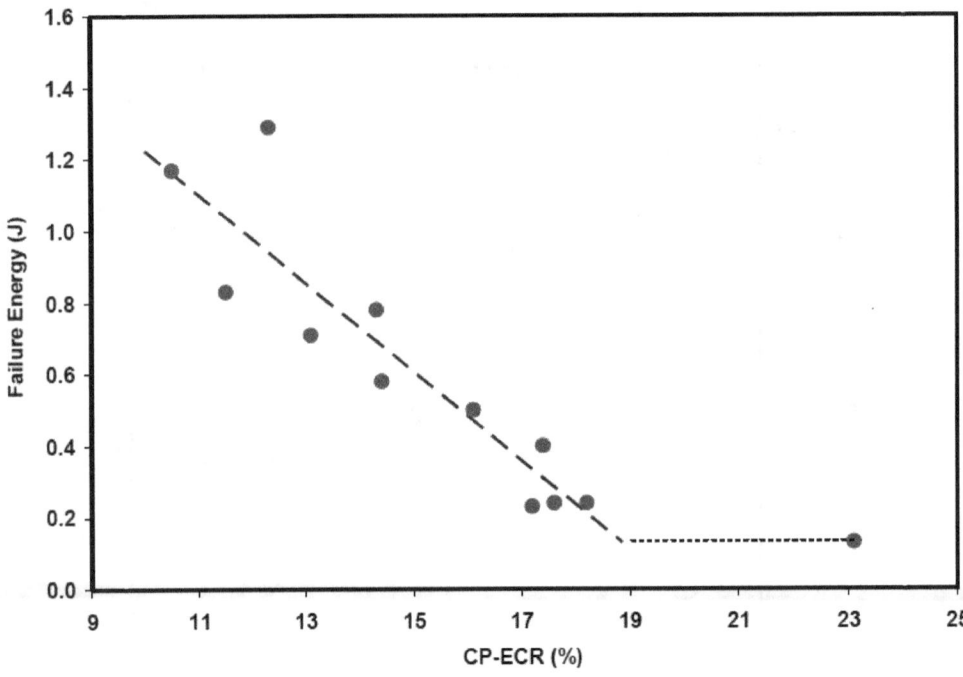

(b) Data for failed samples subjected to 4-PBTs at 135°C

Figure 21. Maximum (for 0% CP-ECR) and failure (for ≥10% CP-ECR) energy as a function of oxidation level (CP-ECR) for post-LOCA samples subjected to 4-PBTs with the rupture region in tension for all tests but one. 4-PBTs were performed at 135°C and 2 or 1 mm/s to 14-mm maximum displacement. One bend test was conducted at 30°C.

38

Offset displacement is a measure of plastic deformation. For three ramp-to-rupture samples (0% CP-ECR), there was a smooth transition between the constant load-displacement slope in the elastic bending regime and the decreasing slope after initiation of plastic flow in the axial direction (see Fig. 22). Measured offset displacements were in the range of 7 to 8 mm for these ductile samples. The loading stiffness values were reasonably close, but they appeared to increase somewhat with decreasing rupture strain: 120 N/mm for 69% rupture strain, 130 N/mm for 33% rupture strain, and 135 N/mm for 21% rupture strain.

The load-displacement curves for oxidized samples were quite different from the ones for ramp-to-rupture samples. Most samples (9 out of 11 quenched samples) severed with an abrupt load drop during the linear portion of the loading ramp. These samples exhibited no offset displacement prior to crack initiation and propagation through the cross section. Six of the nine brittle samples failed outside the rupture node and the remaining three failed in the rupture node.

The six quenched samples that severed outside the rupture node with zero offset strain all had rupture strains ≤32% and hydrogen contents and oxidation levels at the failure locations that would qualify them as brittle based on RCT results for non-deformed LOCA samples. Metallographic imaging and analysis were performed at one of two severed cross sections for the post-bend OCZL#19 sample, and the hydrogen content was determined from rings sectioned close to the severed cross section. Figure 23 shows results of the post-bend characterization. The primary failure occurred at 24 mm below the rupture mid-span with 530-wppm hydrogen and 12% measured ECR. Based on the RCT results (see Fig. 4), cladding samples with this combination of hydrogen content and oxidation level are expected to fail in a brittle manner with no offset or plastic displacement.

The three quenched samples that severed in the rupture node with zero offset strain had rupture strains ≥40% and oxidation levels of 12, 17, and 18% CP-ECR. Figure 24 shows (a) the hydrogen distribution, (b) the severed sample, and (c) the cross section at the severed location for the OCZL#18 sample. The CP-ECR was only 12% and the hydrogen pickup at the failure location was negligible. For a ring with no rupture-opening flaw and uniform oxidation around the circumference, the sample would have exhibited high ductility (30 to 50%) at this oxidation level. However, the local oxidation for this sample varied from much greater than 35% at the rupture tips (see Fig. 25a) to 10% at the thick back region 180° from the center of the rupture opening (see Fig. 25b). Under tensile bending stress, the crack initiated at the brittle rupture tips and propagated rapidly through the cross section with no indication of offset displacement. Although it required a relatively high bending moment and energy to sever this sample, the "ductile" regions of the cladding did not have enough fracture toughness to blunt the growth of the crack.

Figure 26 shows results of the post-bend characterization of a sample that was tested with the rupture opening in compression (OCZL#12). This sample had an average oxidation of 14% CP-ECR in the rupture node, and the primary failure occurred at 40 mm below the rupture mid-span with 1700-wppm hydrogen and 8% ECR. However, severing of the sample at 33 mm above the rupture mid-span appears to be an equally probable failure location based on hydrogen content and oxidation level. Figure 27 shows the OCZL#15 sample, which was another sample subjected to reverse bending (rupture region under compression), after oxidation to 18% CP-ECR and cooling without quench. The offset strain for this intact sample was 13 mm.

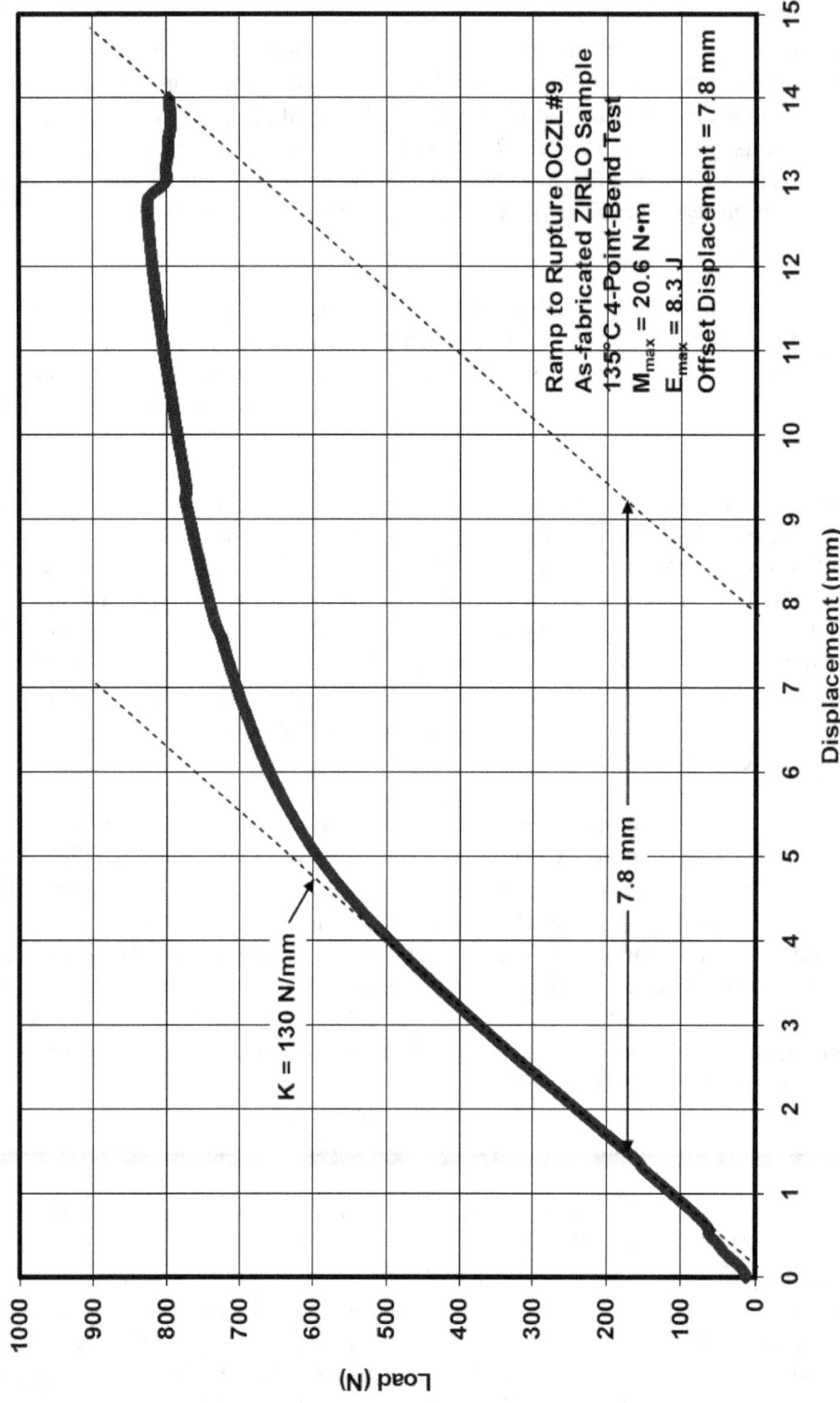

Figure 22. Four-point-bend test (4-PBT) load-displacement curve for ramp-to-rupture sample OCZL#9. The as-fabricated ZIRLO sample ruptured at 875±15°C with a mid-wall rupture strain of 33%. The 4-PBT was conducted at 135°C and 2 mm/s displacement rate to 14-mm maximum displacement.

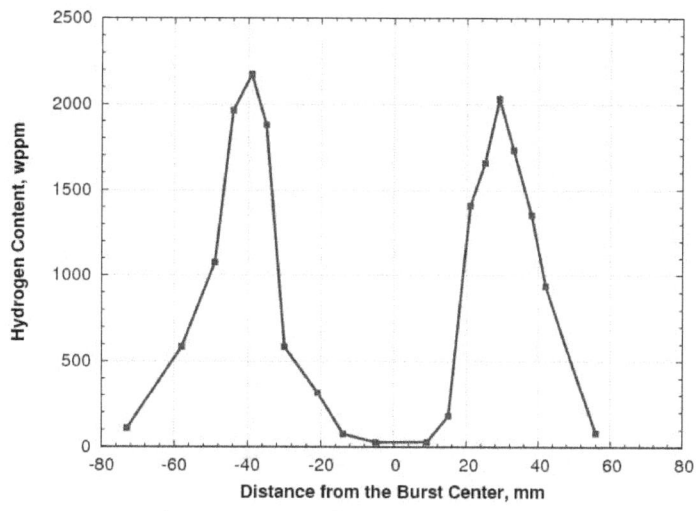

(a) Hydrogen-content profile

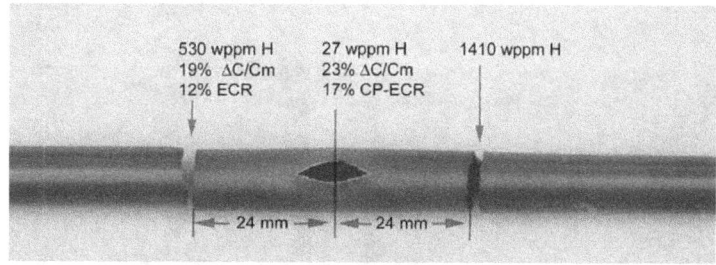

b) Measured values at failure locations

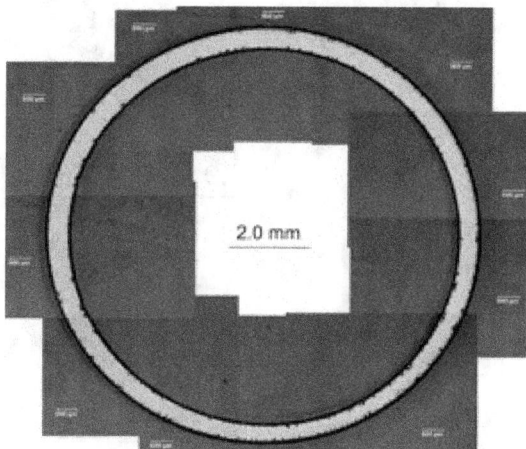

(c) Low-magnification image of severed cross section at -24 mm

Figure 23. Post-bend characterization of OCZL#19 sample that was subjected to bending at 135°C with the rupture region under maximum tensile stress: (a) hydrogen-content profile; (b) measured values at failure locations; and (c) low-magnification image of severed cross section at 24 mm below rupture mid-span.

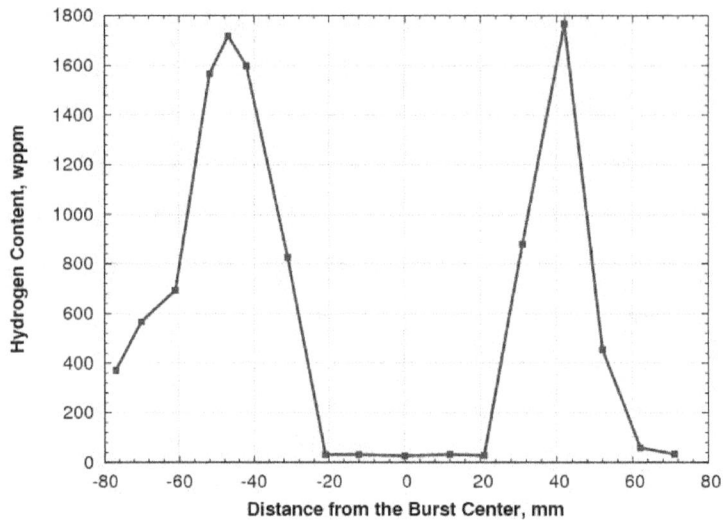

(a) Hydrogen-content profile

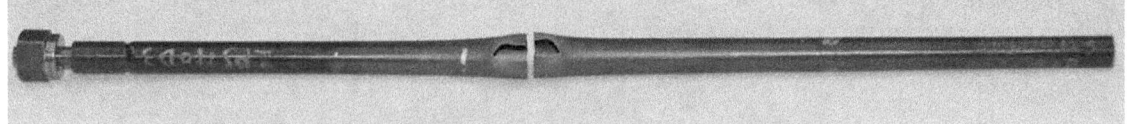

(b) Failure location

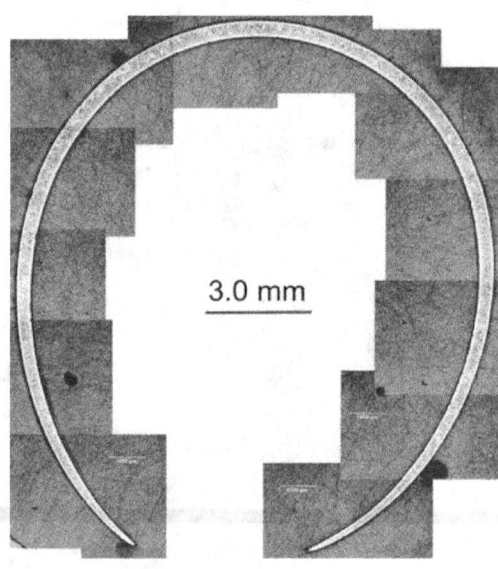

(c) Low-magnification image of severed cross section

Figure 24. Post-bend characterization of the OCZL#18 sample oxidized to 12% CP-ECR, quenched, and subjected to bending at 135°C with the rupture region under maximum tensile stress: (a) hydrogen-content profile; (b) failure location; and (c) low-magnification image of the severed cross section.

42

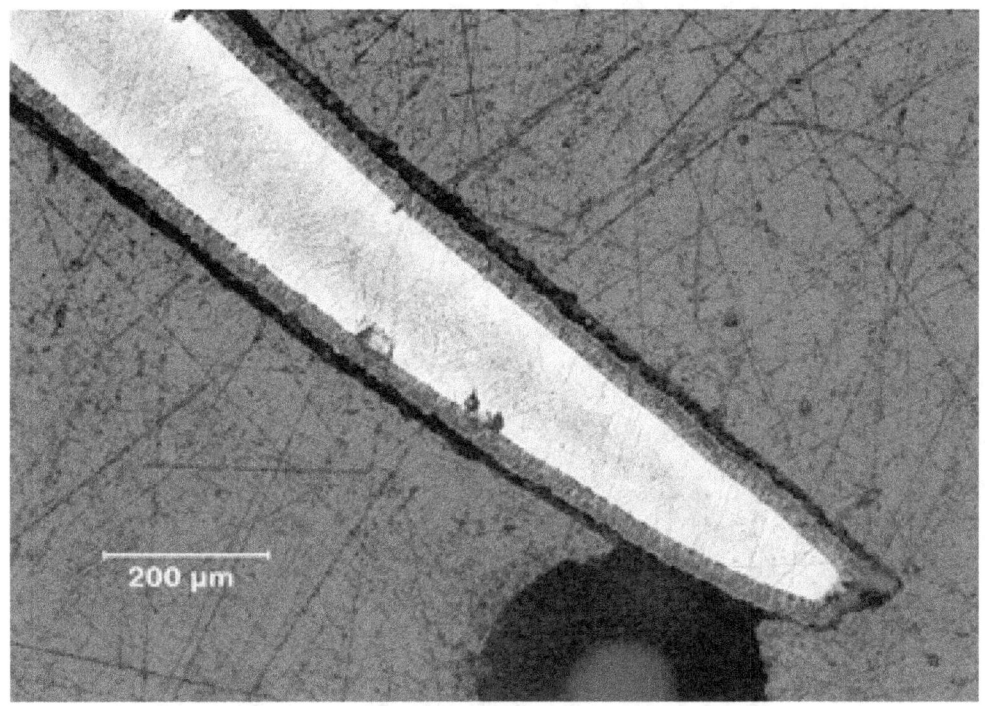

(a) Rupture tip

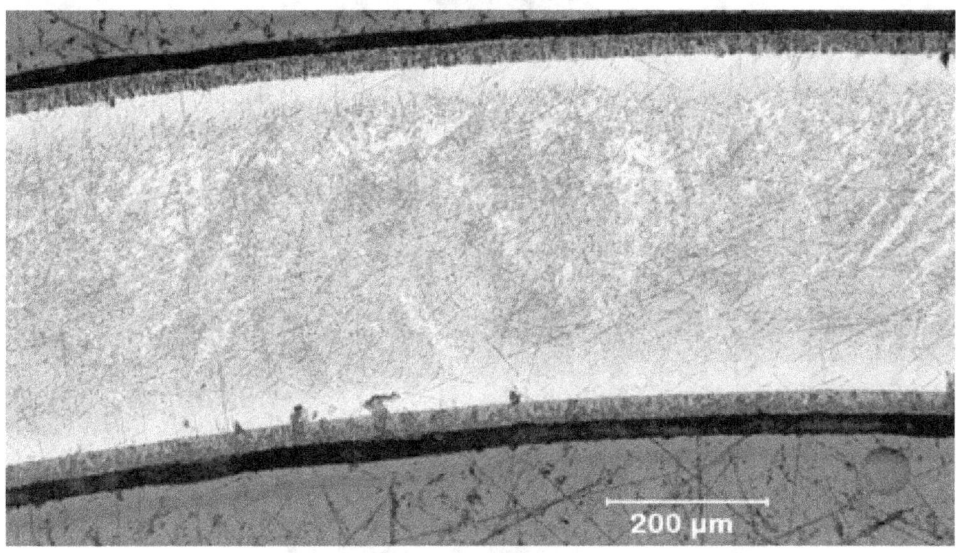

(b) Back side of rupture node cross section

Figure 25. Metallographic images for the OCZL#18 severed cross section following oxidation to 12% CP-ECR, quench, and bending at 135°C with the rupture region under maximum tension: (a) rupture tip with 0.14-mm average metal wall and (b) back side of cross section with 0.44-mm metal wall thickness.

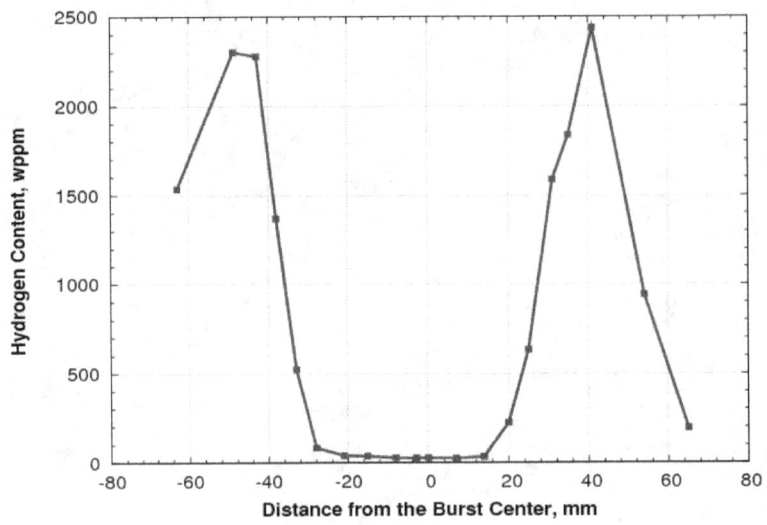

(a) Hydrogen-content profile

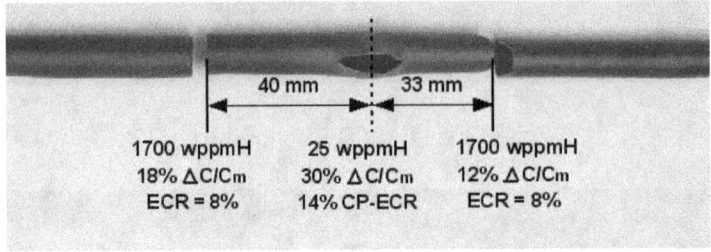

(b) Measured values at failure locations

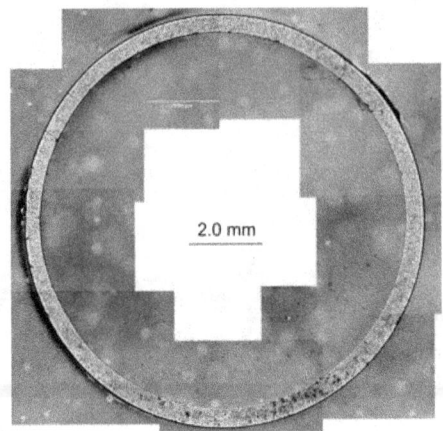

(c) Low-magnification image of severed cross section at -40 mm

Figure 26. Post-bend characterization of OCZL#12 sample that was subjected to bending at 135°C with the rupture region under maximum compressive stress: (a) hydrogen-content profile; (b) measured values at failure locations; and (c) low-magnification image of severed cross section at 40 mm below rupture mid-span.

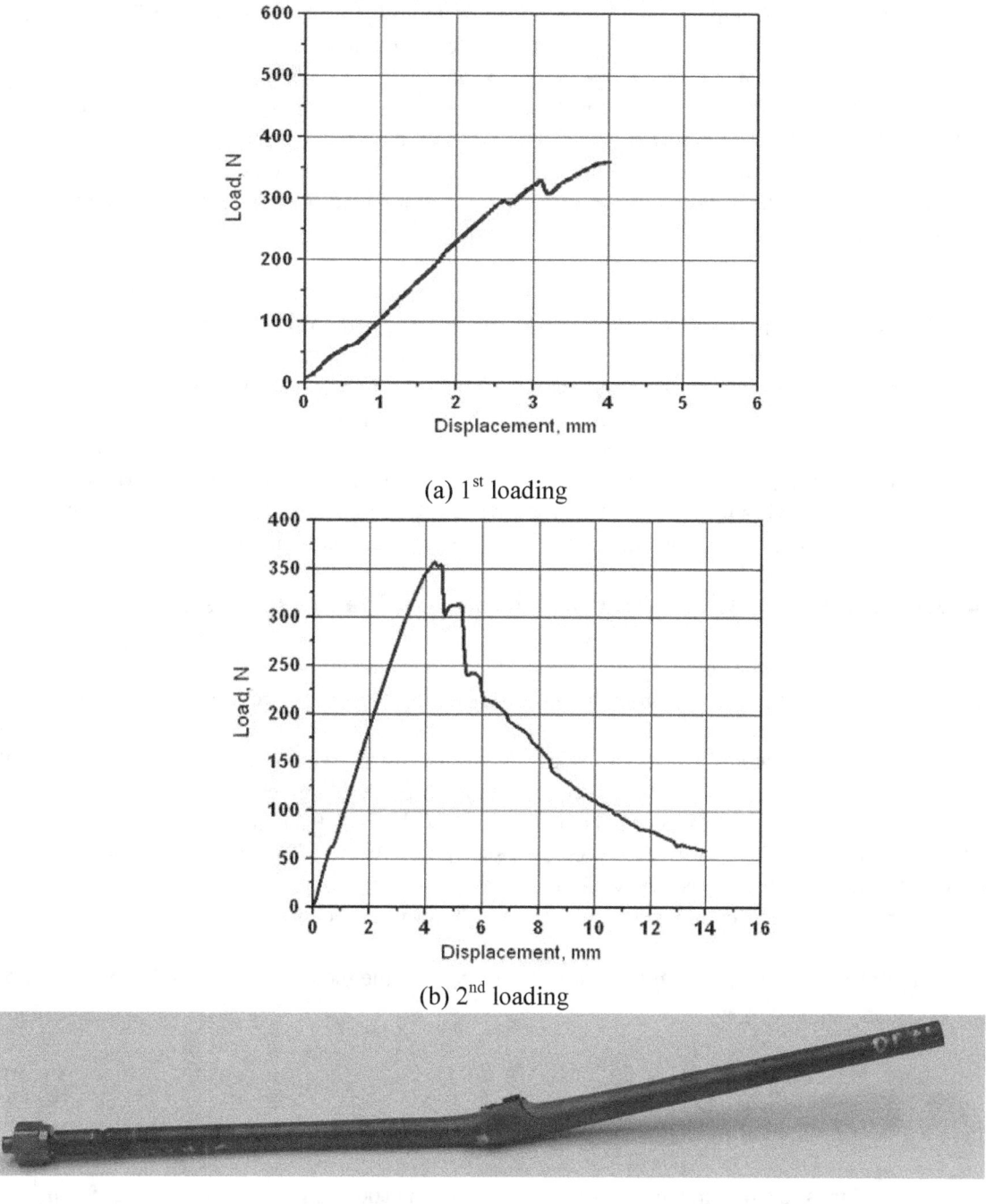

(a) 1st loading

(b) 2nd loading

(c) Sample appearance after 2nd loading

Figure 27. Post-bend results for OCZL#15 sample (18% CP-ECR and quench) with rupture node in compression (reverse bending), 135°C test temperature and 2 mm/s displacement rate: (a) load-displacement curve for 1st loading; (b) load-displacement curve for 2nd loading; and (c) sample appearance after 2nd loading.

Figure 28 shows the OCZL#17 sample, which was subjected to standard bending (rupture region under tension). The back region of the cladding remained intact after 10-mm displacement with an offset displacement of 6.3 mm. In Fig. 28a, the load-displacement curve is truncated at 4-mm displacement.

From 4- to 10-mm displacement, the load remained quite low (3 to 13 N) as the ductile ligament at the back of the cladding deformed plastically. This ductile deformation region contributed about 10% to the maximum energy. The OCZL#29 sample was also subjected to the standard ANL bend test. This test was terminated after ≈3-mm displacement (Fig. 29) and reloaded (Fig. 30). Cracking began at a relatively low load (188 N) and bending moment (4.7 N•m). The load dropped rather abruptly from 188 N to 107 N and again from 124 N to 85 N. At the end of these two load drops, the crack had propagated across more than half the cross section. After reloading, the crack continued to propagate in a ductile manner leaving only the back region intact.

In summary, the post-LOCA 4-PBT load-displacement curves give three types of results: severing with zero offset displacement in brittle regions outside the rupture node; severing with zero offset displacement in brittle-to-ductile regions of the rupture cross section; and partial-wall, brittle cracking followed by ductile crack growth during which offset strain is observed. It is clear from these tests at ≤18% CP-ECR that a significant fraction of the cladding cross section in the rupture node has plasticity; yet most of these tests show no offset displacement. Offset displacement appears to be affected by balloon size, loading history, circumferential temperature gradient, and other factors not yet quantified. Further, well behaved trends are not observed as they are for bending moment and failure energy. Although offset displacement is related to ductility, it is not a useful metric for behavior of the balloon region.

3.4 Comparison of ANL 4-PBT and JAEA Axial-restraint Tests

ANL 4-PBT results can be compared to JAEA axial-restraint results as both tests induce axial stresses. The stress distributions are different in that the stress acting on a cross section is tensile everywhere in the axial tensile test, while the stress in the axial bend tests varies from tensile to compressive across the cross section. However, failure of a brittle material is governed by the maximum tensile stress. Given the failure bending moment for a 4-PBT, the equivalent axial force that gives the maximum tensile stress can be determined. This determination can be made more easily for 4-PBT samples that fail outside the rupture region. Figure 23c shows the severed cross section for the 17% CP-ECR sample from Test OCZL#19. The cross section is nearly circular. Table 6 gives the dimensions determined from profilometry, low-magnification metallography, and high-magnification metallography.

For the 4-PBT sample, the maximum tensile stress (σ_{max}) in the metal is related to the maximum bending moment (M_{max}) according to:

$$\sigma_{max} = M_{max} R_{mo}/I_{eq} \tag{7}$$

where R_{mo} is the outer radius ($R_{mo} = D_{mo}/2$) of the metal wall (5.285 mm in this case).

The maximum tensile stress in the metal is σ_{max} = 5.7 N•m (1000-mm/m) (5.285 mm)/248 mm^4 = 121.5 MPa.

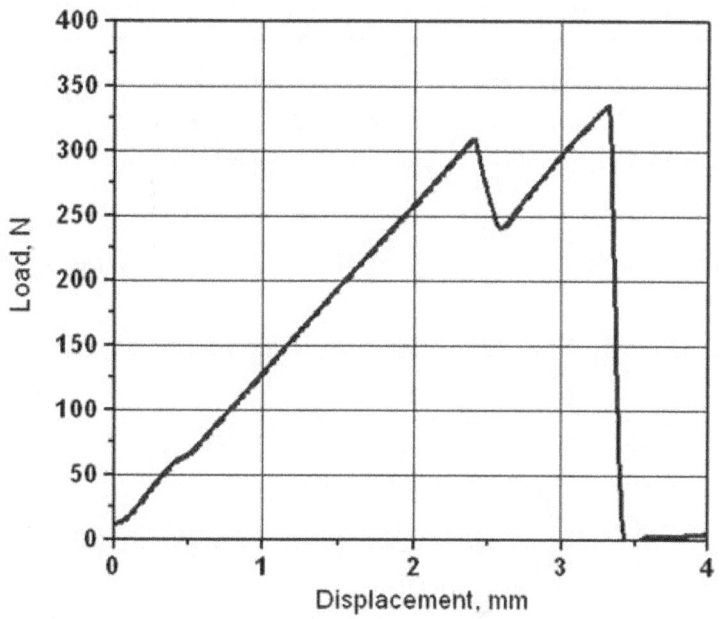

(a) Load-displacement curve for 1st 4-mm displacement

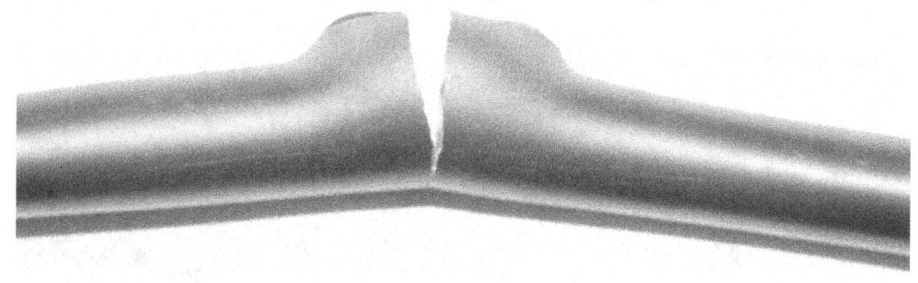

(b) Post-test image of side view

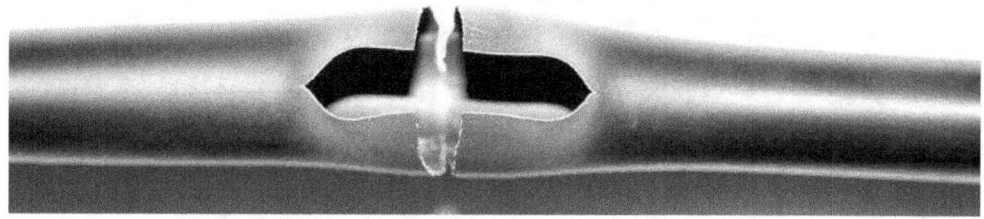

(c) Post-test image of rupture view

Figure 28. 4-PBT results for Test Sample OCZL#17 oxidized to 13% CP-ECR, quenched, and tested at 135°C and 2 mm/s displacement rate: (a) load-displacement curve for 1st 4-mm displacement (out of 10-mm total displacement); (b) post-test image of side view; and (c) post-test image of rupture view.

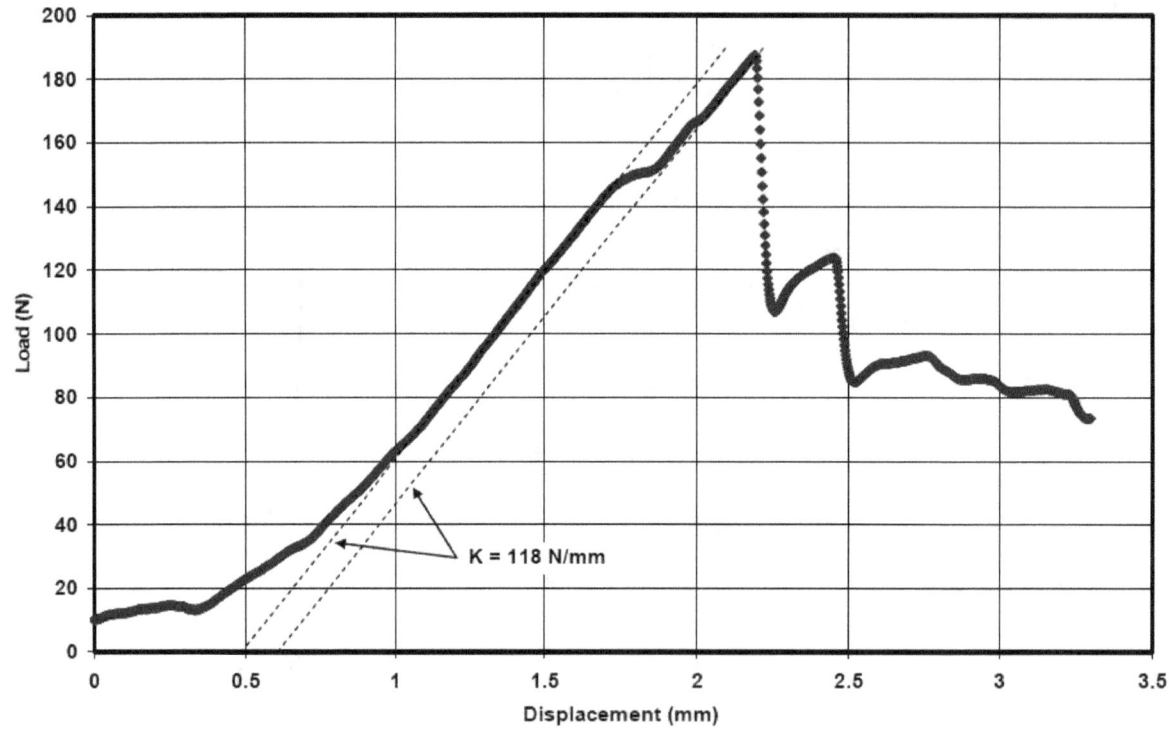

(a) 1st loading-unloading sequence

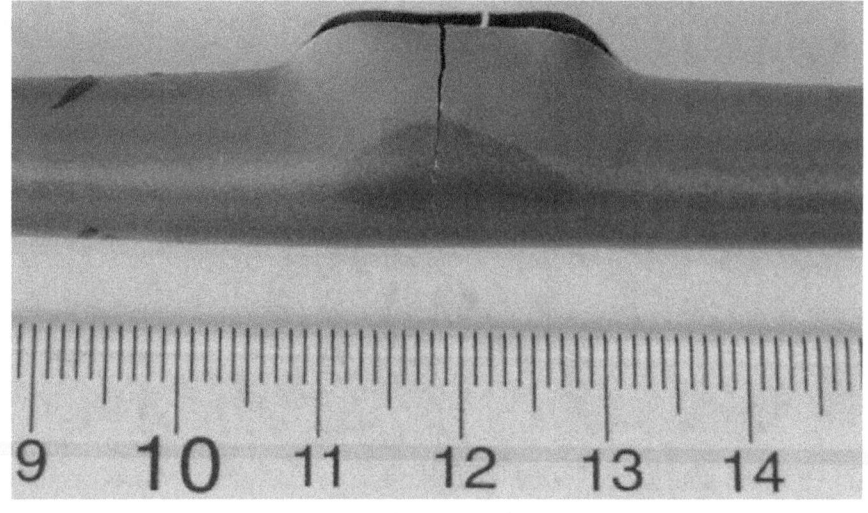

(b) Side view after 1st loading-unloading

Figure 29. Test Sample OCZL#29 oxidized to 17% CP-ECR, quenched, and subjected to 4-PBTs at 135°C and 1 mm/s displacement rate: (a) load-displacement curve for 1st loading-unloading sequence and (b) sample appearance after 1st loading-unloading.

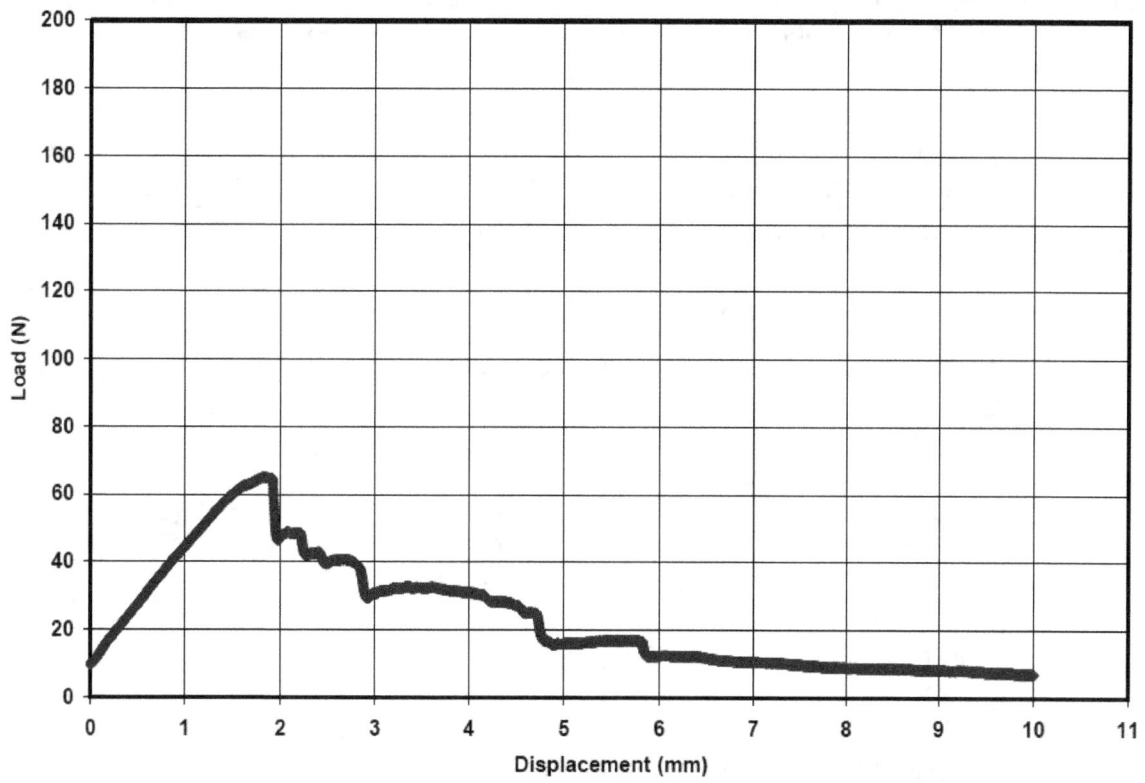

(a) Second loading-unloading

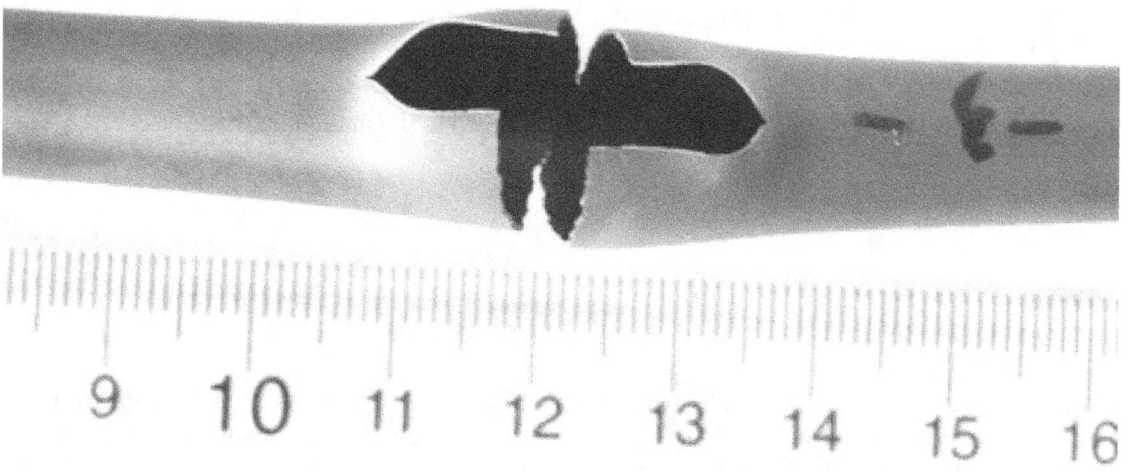

(b) Rupture view after 2nd loading-unloading

Figure 30. Test Sample OCZL#29 oxidized to 17% CP-ECR, quenched, and subjected to 4-PBTs at 135°C and 1 mm/s displacement rate: (a) load-displacement curve for 2nd loading-unloading sequence and (b) sample appearance after 2nd loading-unloading.

Table 6. Dimensions, properties, and failure bending moment at severed location for test sample OCZL#19 with 17% maximum CP-ECR in the rupture node and 12% CP-ECR at the severed location.

Parameter	Value	Comment
Outer Diameter (OD, D_o), mm	10.67	Measured
OD Oxide Thickness $(\delta_{ox})_o$, μm	50±6	Measured
Metal OD (D_{mo}), mm	10.57	$D_{mo} = D_o - 2\,(\delta_{ox})_o$
Metal Wall Thickness (h_m), mm	0.46±0.04	Measured
Metal ID (D_{mi}), mm	9.65	$D_{mi} = D_{mo} - 2\,h_m$
ID Oxide Thickness $(\delta_{ox})_i$, μm	35±6	Measured
Inner Diameter (ID, D_i), mm	9.58	$D_i = D_{mi} - 2\,(\delta_{ox})_i$
Cross-sectional Area (A), mm^2 OD Oxide $(A_{ox})_o$ Metal (A_m) ID Oxide $(A_{ox})_i$	 1.67 14.61 1.06	 $(A_{ox})_o = (\pi/4)\,[(D_o)^2 - (D_{mo})^2]$ $A_m = (\pi/4)\,[(D_{mo})^2 - (D_{mi})^2]$ $(A_{ox})_i = (\pi/4)\,[(D_{mi})^2 - (D_i)^2]$
Area Moment of Inertia (I), mm^4 OD Oxide $(I_{ox})_o$ Metal (I_m) ID Oxide $(I_{ox})_i$	 23.5 187.1 12.2	 $(I_{ox})_o = (\pi/64)\,[(D_o)^4 - (D_{mo})^4]$ $I_m = (\pi/64)\,[(D_{mo})^4 - (D_{mi})^4]$ $(I_{ox})_i = (\pi/64)\,[(D_{mi})^4 - (D_i)^4]$
Young's Modulus (E) at 135°C, MPa Oxide (E_{ox}) Metal (E_m)	 14.8×10^4 8.65×10^4	 MATPRO [3] MATPRO [3]
Equivalent A (A_{eq}) Relative to Metal, mm^2	19.3	$A_{eq} = A_m + (E_{ox}/E_m)\,A_{ox}$
Equivalent I (I_{eq}) Relative to Metal, mm^4	248	$I_{eq} = I_m + (E_{ox}/E_m)\,I_{ox}$
Maximum Bending Moment (M_{max}), N•m	5.7	Measured

For the axial tensile test, the maximum load (P_{max}) is related to the maximum metal stress (σ_{max}) according to:

$$P_{max} = \sigma_{max}\,A_{eq} \tag{8}$$

Setting σ_{max} = 121.5 MPa and A_{eq} = 19.3 mm^2 gives P_{max} = 2345 N. If the correlation value (see Eq. 5) of 6.4 N•m at 17% CP-ECR had been used, then P_{max} would be equal to 2632 N.

Based on the results for the JAEA fully restrained samples [13,14], the maximum load measured for samples that survived quench from 700°C to <100°C was in the range of 1200 N to 2400 N. For samples oxidized at 1200°C to >12% BJ-ECR, the maximum was 1200 to 2000 N. These results are for cladding with pre-hydride levels ≥ 100 wppm. Thus, even under these conditions, it appears that the ANL sample from Test OCZL#19 would have survived JAEA fully restrained quench without failure. As the ANL bend tests were performed at 135°C, it would be more proper to say that the 17% CP-ECR OCZL#19 sample would have survived the JAEA full-constraint test with quench from 700°C to 135°C.

A similar analysis was performed for the OCZL#12 sample which had 14% maximum CP-ECR in the rupture node and 8% measured ECR at the severed location 40 mm below the rupture mid-span. The failure bending moment was 10.5 N•m and the equivalent cross-sectional area and area moment of inertia were 17.9 mm^2 and 249 mm^4, respectively. Using these values, along with a metal outer radius of 5.467 mm, gives an equivalent axial failure load of 4127 N, which is much higher than any load measured in the JAEA fully constrained tests.

Equations 7 and 8 can be combined to give the JAEA tensile failure load (P_{max}) as a function of the ANL failure bending moment (M_{max}):

$$P_{max} = M_{max} R_{mo} (A_{eq}/I_{eq}) \tag{9}$$

In principle, Eq. 9 can be used for samples that severed in the rupture node (e.g., OCZL#29 at 17% CP-ECR). However, the severed cross section for the OCZL#29 sample (49% rupture strain) would be similar to the one shown in Fig. 24c for the OCZL#18 sample (43% rupture strain). The equivalent area and area moment of inertia would have to be calculated numerically for this case. Although the maximum bending moment for the OCZL#29 sample (4.7 N•m) was less than for the OCZL#19 sample (5.7 N•m), the sample did exhibit plastic flow following the first significant load drop and did not sever into two pieces. This adds additional complexity to the determination of the axial load needed to sever the sample at this location.

4 CONCLUSIONS

The LOCA acceptance criteria that limit peak oxidation temperature and maximum oxidation level are based on retention of ductility. For non-deformed (i.e., non-ballooned) regions of a fuel rod during a LOCA, the ductility concept is unambiguous and can be determined with straightforward testing. Among the testing methods evaluated, 3-PTB tests and RCTs are quite good for LOCA evaluation because they can measure loads and displacements readily – and they give similar results with regard to the ductile-to-brittle transition oxidation level as a function of pre-oxidation hydrogen content. The 3-PBT has advantages for as-fabricated and pre-hydrided material when long specimens are available. However, RCTs use much shorter specimens, and this becomes advantageous when testing irradiated material. Therefore, for comprehensive studies of as-fabricated, pre-hydrided, and irradiated cladding, the RCT is preferred.

Although the LOCA temperature and oxidation criteria may also protect ballooned and ruptured regions from severing and fragmenting during and following quench, ductility will not be retained everywhere in this region and interpretation of test results is not straightforward. Higher hydrogen-content regions from the rupture edge to the hydrogen peak will contain brittle material. Also, within the rupture region cross sections, the cladding transitions from brittle (at the thin, heavily oxidized rupture tips) to ductile (at the thicker back region which is at lower oxidation level) at the averaged 17% oxidation level.

Of the two methods being used for testing ballooned sections, the ANL 4-PBT is preferred over the JAEA full- or partial-restraint quench test because load-versus-displacement curves can be measured accurately in the 4-PBT. This permits the determination of offset displacement (measure of plastic deformation), maximum bending moment (measure of strength), and failure energy (measure of toughness). Offset displacement of a ballooned segment is not well behaved, however, and it is therefore not a useful metric for structural behavior. No offset displacement was observed for bending failures in the region bounded from the rupture tip to the hydrogen peak near the neck of the balloon. The offset displacement was also zero prior to >50% severing of the cross section within the rupture region.

On the other hand, the 4-PBT maximum bending moments and failure energies exhibited smooth trends toward very low values with increasing oxidation level up to 19% CP-ECR, but remained significantly above zero for oxidation levels ≤17%. Our calculation based on bending moment showed that a sample oxidized to 17% CP-ECR would have sufficient strength to survive the fully restrained tests performed by JAEA. Further, none of our unrestrained samples failed upon quench, and none of the samples tested to failure in the ANL program fragmented or failed in a "low-toughness" mode. By contrast, glass and ceramic rods with much lower toughness than the post-LOCA samples sever into as many as 5 to 10 pieces when subjected to 4-PBT loading. Therefore, the 4-PBT can discriminate between good and poor structural performance of ballooned segments. The ANL data used for this evaluation show that the current limits of 17% CP-ECR and 1204°C protect fresh cladding during quench not only from fragmentation but also from severing into two pieces under a wide range of loading conditions.

REFERENCES

1. Michael Billone, Yong Yan, Tatiana Burtseva, and Robert Daum, "Cladding Embrittlement during Postulated Loss-of-Coolant Accidents," NUREG/CR-6967, July, 2008; available online in as ML082130389 at http://www.nrc.gov/reading-rm/adams.html.

2. Y. Yan, T. A. Burtseva, and M. C. Billone, "Post-quench Ductility Results for North Anna High-burnup 17×17 ZIRLO Cladding with Intermediate Hydrogen Content," ANL letter report to NRC, April. 17, 2009; available online as ML091200702 at http://www.nrc.gov/reading-rm/adams.html.

3. L.J. Siefken, E.W. Coryell, E.A. Harvego, and J.K. Hohorst, *MATPRO – A Library of Materials Properties for Light-Water-Reactor Accident Analysis*, in SCDAP/RELAP5/MOD 3.3 Code Manual, NUREG/CR-6150, Vol. 4, Rev. 2, available as ML010330424 at http://www.nrc.gov/reading-rm/adams.html.

4. M. C. Billone, Y. Yan, and T. A. Burtseva, "Procedure for Conducting Oxidation and Post-Quench Ductility Tests with Zirconium-based Cladding Alloys," ANL report to NRC, Mar. 31, 2009; available online as ML090900841 at http://www.nrc.gov/reading-rm/adams.html.

5. J.-C. Brachet, V. Vandenberghe-Maillot, L. Portier, D. Gilbon, A. Lesbros, N. Waeckel, and J.-P. Mardon, "Hydrogen Content, Preoxidation, and Cooling Scenario Effects on Post-Quench Microstructure and Mechanical Properties of Zircaloy-4 and M5® Alloys in LOCA Conditions," J. ASTM Intl., Vol. 5, No. 5 (2008). Available online as JAI101116 at http://www.astm.org.

6. V. Vandenberghe, J.C. Brachet, M. Le Saux, D. Gilbon, M. Billone, D. Hamon, J.P. Mardon and H. Hafidi, "Influence of the Cooling Scenario on the Post-Quench Mechanical Properties of Pre-Hydrided Zircaloy-4 Fuel Claddings after High Temperature Steam Oxidation (LOCA Conditions)," Proc. 2010 LWR Fuel Performance/TopFuel/WRFPM, Orlando, FL, Sept. 26-29, 2010, Paper 096.

7. R. S. Daum, S. Majumdar, H. Tsai, T. S. Bray, D. A. Koss, A. T. Motta, and M. C. Billone, "Mechanical Property Testing of Irradiated Zircaloy Cladding under Reactor Transient Conditions," *Small Specimen Test Techniques: Fourth Volume, ASTM STP 1418*, M. A. Sokolov, J. D. Landes, and G. E. Lucas, Eds., American Society for Testing and Materials, West Conshohocken, PA, 2002.

8. R. A Jaramillo, W. R. Hendrich, and N. H. Packan, "Tensile Hoop Behavior of Irradiated Zircaloy-4 Nuclear Fuel Cladding," ORNL/TM-2007/012, March 2007. Available online at http://www.osti.gov/bridge/product.biblio.jsp?query_id=0&page=0&osti_id=931509&Row=0.

9. V. Grigoriev, R. Jacobsson, and D. Schrire, "Temperature Effect on BWR Cladding Failure under Mechanically Simulated RIA Conditions," JAERI – Conf 2002 – 009, FSRM Procs., March 4-5, 2002, Tokai, pp. 97-106; available online at http://sciencelinks.jp/j-east/article/200223/000020022302A0845597.php.

10. Y. Yan, T.A. Burtseva, and M.C. Billone, "Post-LOCA Bend Test Results for As-fabricated Cladding," ANL letter report to NRC, Feb. 9, 2010; available online as ML100470686 at http://www.nrc.gov/reading-rm/adams.html.

11. Y. Yan, T.A. Burtseva, R.O. Meyer, and M.C. Billone, "Update of LOCA-Integral and Post-LOCA-Bend Test Results for Fresh ZIRLO Cladding," ANL letter report to NRC, July 21, 2010; available online as ML111380437 at http://www.nrc.gov/reading-rm/adams.html.

12. Y. Yan, T.A. Burtseva, R.O. Meyer, and M.C. Billone, "Argonne Results for ANL-Studsvik Benchmark Tests," ANL letter report to NRC, Aug. 13, 2010; available online as ML111380445 at http://www.nrc.gov/reading-rm/adams.html.

13. Fumihisa Nagase and Toyoshi Fuketa, "Effect of Pre-Hydriding on Thermal Shock Resistance of Zircaloy-4 Cladding under Simulated Loss-of-Coolant Accident Conditions," JNST, Vol. 41, No. 7 (2004) 723-730.

14. Fumihisa Nagase and Toyoshi Fuketa, "Behavior of Pre-hydrided Zircaloy-4 Cladding under Simulated LOCA Conditions," JNST, Vol. 42, No. 2 (2005) 209-218.

15. Fumihisa Nagase and Toyoshi Fuketa, "Fracture Behavior of Irradiated Zircaloy-4 Cladding under Simulated LOCA Conditions," JNST, Vol. 43, No. 9 (2006) 1114-1119.

16. Fumihisa Nagase, Toshinori Chuto, and Toyoshi Fuketa, "Behavior of High Burn-up Fuel Cladding under LOCA Conditions," JNST, Vol. 46, No. 7 (2009) 763-769.

17. F. Nagase and T. Fuketa, "Thermal Shock Resistance of Irradiated PWR Claddings," *SEGFSM Topical Meeting on LOCA Fuel Issues*, NEA/CSNI/R(2004)19, November 2004, 243-256.

NRC FORM 335 (12-2010) NRCMD 3.7	U.S. NUCLEAR REGULATORY COMMISSION **BIBLIOGRAPHIC DATA SHEET** *(See instructions on the reverse)*	1. REPORT NUMBER (Assigned by NRC, Add Vol., Supp., Rev., and Addendum Numbers, if any.) NUREG/CR-7139

2. TITLE AND SUBTITLE Assessment of Current Test Methods for Post-LOCA Cladding Behavior	3. DATE REPORT PUBLISHED	
	MONTH	YEAR
	August	2012
	4. FIN OR GRANT NUMBER JCN V6199	

5. AUTHOR(S) M. C. Billone	6. TYPE OF REPORT Technical
	7. PERIOD COVERED (Inclusive Dates) 2009-2011

8. PERFORMING ORGANIZATION - NAME AND ADDRESS (If NRC, provide Division, Office or Region, U. S. Nuclear Regulatory Commission, and mailing address; if contractor, provide name and mailing address.)

Argonne National Laboratory
9700 South Cass Ave
Argonne, Illinois 60439

9. SPONSORING ORGANIZATION - NAME AND ADDRESS (If NRC, type "Same as above", if contractor, provide NRC Division, Office or Region, U. S. Nuclear Regulatory Commission, and mailing address.)

Division of Systems Analysis
Office of Nuclear Regulatory Research
U.S. Nuclear Regulatory Commission
Washington, D.C. 20555-0001

10. SUPPLEMENTARY NOTES

11. ABSTRACT (200 words or less)

Test methods to assess fuel-rod cladding behavior following a loss-of-coolant accident (LOCA) are compared and evaluated. For irradiated cladding, the ring compression test (RCT) is the best test method for generating ductility data for assessing the effects of irradiation and hydrogen pickup on embrittlement oxidation threshold. However, neither 3-PBTs nor RCTs are useful for evaluating the performance of ballooned and ruptured cladding with significant axial gradients in cladding geometry, oxidation level, and hydrogen content, as well as circumferential gradients in wall thickness and oxidation level within the rupture region. Partially restrained axial contraction tests are useful for determining the fracture/nofracture boundary for ballooned, ruptured, oxidized, and quenched cladding as a function of hydrogen content and oxidation level. The four-point bend test (4-PBT) is best for determining three cladding performance metrics: maximum bending moment (measure of strength), failure energy (measure of toughness), and offset displacement (measure of plastic deformation).

12. KEY WORDS/DESCRIPTORS (List words or phrases that will assist researchers in locating the report.) LOCA test methods zirconium alloy cladding oxidation and embrittlement bending moment	13. AVAILABILITY STATEMENT unlimited
	14. SECURITY CLASSIFICATION
	(This Page) unclassified
	(This Report) unclassified
	15. NUMBER OF PAGES
	16. PRICE

NRC FORM 335 (12-2010)

UNITED STATES
NUCLEAR REGULATORY COMMISSION
WASHINGTON, DC 20555-0001

OFFICIAL BUSINESS

NUREG/CR-7139

Assessment of Current Test Methods for Post-LOCA Cladding Behavior

August 2012

www.ingramcontent.com/pod-product-compliance
Lightning Source LLC
Chambersburg PA
CBHW081836170526
45167CB00007B/2834